ADVANCED ALGEBRA

ADVANCED ALGEBRA

BY

E. A. MAXWELL, Ph.D.

Fellow of Queens' College
Cambridge

PART I

CAMBRIDGE
AT THE UNIVERSITY PRESS
1966

CAMBRIDGE UNIVERSITY PRESS
Cambridge, New York, Melbourne, Madrid, Cape Town, Singapore, São Paulo, Delhi

Cambridge University Press
The Edinburgh Building, Cambridge CB2 8RU, UK

Published in the United States of America by Cambridge University Press, New York

www.cambridge.org
Information on this title: www.cambridge.org/9780521102674

© Cambridge University Press 1966

This publication is in copyright. Subject to statutory exception and to the provisions of relevant collective licensing agreements, no reproduction of any part may take place without the written permission of Cambridge University Press.

First published 1960
Reprinted 1961, 1962, 1966
This digitally printed version 2009

A catalogue record for this publication is available from the British Library

ISBN 978-0-521-05694-6 hardback
ISBN 978-0-521-10267-4 paperback

CONTENTS

PREFACE

I am most grateful to Mr A. P. Rollett for reading the manuscript and making a number of valuable suggestions; also to two practising schoolmasters who read it on behalf of the Press and put forward some very useful comments. The Examples are taken chiefly from papers set in the University of Cambridge and by the Oxford and Cambridge Schools Examination Board; I am grateful for permission to use them, and also for help given by my son in checking answers.

The Staff of the Cambridge University Press have, as always, combined patience and efficiency in the passage from manuscript to book, and I am deeply grateful to them.

I mention here that I share the royalties with Queens' College and with Holy Trinity Church, both of Cambridge, so that I may record my gratitude for all that they have meant to me over many years.

E. A. M.

June 1959

One or two small corrections have been made during reprinting. In particular, Illustration 9 on p. 189 is now correct.

E. A. M.

May 1961

Persistent, but kindly, correspondents have drawn attention to a number of errors, particularly in answers, and I hope that corrections are now virtually complete.

E.A.M.

January 1965

INTRODUCTION

My aims in the preparation of this book have been two-fold: to emphasize the logical structure of algebra up to the limits, as I judge them, of the normal reader's capacity for this stage, and to develop a running technique which will enable problems to be tackled with reasonable fluency. I make no apology for the large number of routine examples—only sympathy from one who has himself worked them all. It is my firm conviction that mathematics, like music with which it is often associated, cannot yield its full thrills to those who will not endure practice.

The subject-matter is necessarily fairly standard, and ought to form a basis for most upper-school requirements below full Scholarship level, and, in places, beyond. One or two points of detail ought perhaps to be mentioned:

I believe the treatment of partial fractions to be new, particularly for quadratic factors of the denominator. The identification of the numerator (of type $Ax+B$) allows justification without recourse to polynomial theory.

It is doubtful whether the exponential and logarithmic series belong properly to algebra, especially as they are almost always included now in the calculus course. The present treatment is a variant of that in, say, my *An Analytical Calculus* and is intended to emphasize the more algebraic aspects of the subjects. I have not hestitated to use elementary calculus when necessary. I hope that the short chapter on Infinite Series will help to prepare a basis for the binomial, exponential and logarithmic series which follow.

The treatment of determinants is fairly orthodox, but I have tried to fix attention on those parts of the work which form a preparation for the important subject of 'linear algebra'.

It is my hope that the reader will be able to pass fully prepared to the next stages unencumbered with material for which he will have no further use. But this is a goal that no author dare feel confident of reaching.

1

SOME FUNDAMENTAL IDEAS AND NOTATIONS

1. 'Knowns' and 'unknowns'; constants and variables. When a problem is stated in the language of algebra, there are usually certain numbers regarded as given and others whose values have to be found. The former are *constants* of the problem, and may be ordinary arithmetical numbers, or algebraic numbers denoted by symbols $a, b, c, d, \ldots$. The latter are often, but by no means always, denoted by letters like x, y, z from the end of the alphabet. In this context they are *unknowns*; but they are also called *variables* since they are, at any rate to begin with, capable of assuming varying values until finally pinned down by the conditions of the problem.

Illustration 1. *To determine two numbers whose sum is* $3a$ *and the sum of whose squares is* $5a^2$.

The statement of the problem involves the two arithmetical constants 3, 5 and the algebraic constant a. To solve it, assume that the 'unknowns' have values x, y. The first condition imposes the restriction

$$x+y = 3a,$$

and the second the restriction

$$x^2+y^2 = 5a^2.$$

If the first equation is considered in isolation, then x, y can take an infinite variety of values, for example $0, 3a$; $a, 2a$; $2a, a$; $3a, 0$; $4a, -a$; $5a, -2a$; and so on. Thus x, y appear as genuine variables. But when the second equation is superposed, the values of x, y are restricted very completely. For, by the first equation,

$$x = 3a-y,$$

so that, by the second equation,

$$(3a-y)^2+y^2 = 5a^2,$$

or

$$2y^2-6ay+4a^2 = 0,$$

or

$$2(y-a)(y-2a) = 0.$$

Hence *either* $y = a$, so that $x = 3a-a = 2a$,

or $y = 2a$, so that $x = 3a-2a = a$.

The solutions of the two equations are thus *either* $x = 2a, y = a$, *or* $x = a, y = 2a$.

What appeared as variables were in fact constants all the time; but the conception of them as variables *in the first instance* was essential to the solution of the problem.

Note for the more experienced reader. This is a convenient point for a short logical note which the beginner may prefer to postpone. What we have actually proved is that, *if the two equations*

$$x+y = 3a,$$
$$x^2+y^2 = 5a^2$$

have any solutions at all, then they are either $x = 2a$, $y = a$ or $x = a$, $y = 2a$. It is, logically, not yet established that the equations are in fact soluble, and the argument ought to be completed by substituting the available values in the given equation, when all is well.

In this case, substitution shows at once that both solutions are possible, and the problem is then solved absolutely. There are, however, examples in which this is not so. Suppose, say, that x is required to satisfy the equation

$$1+\sqrt{x} = \sqrt{(5-x)},$$

where *positive* square roots are to be taken. Since the two sides are equal, so also are their squares, so that

$$1+2\sqrt{x}+x = 5-x,$$

or

$$\sqrt{x} = 2-x.$$

Once again, the squares are equal, so that

$$x = 4-4x+x^2,$$

or

$$x^2-5x+4 = 0,$$

or

$$(x-1)(x-4) = 0.$$

Hence x is *either* 1 *or* 4. But verification in the original equation shows that $x = 1$ is a solution whereas $x = 4$ is not.

In practice, the final verification is often (indeed, usually) omitted, but care must be taken in doubtful cases, especially when square roots are involved. See the report *The Teaching of Algebra in Sixth Forms* issued by the Mathematical Association, 1957, where chapter II gives an excellent account of the problem.

Warning Example. Criticize the following solution of the equation

$$x+1 = 5:$$

'Square each side. Hence $(x+1)^2 = 25$, so that $x^2+2x-24 = 0$, or $(x-4)(x+6) = 0$. Thus $x = 4$ or $x = -6$.'

2. Functions. An isolated expression like

$$3^3+4^3+5^3$$

has the surprising property that its value is 6^3, but, in the context of 'constants' and 'variables', has little further interest. On the other hand, the expression

$$(x-1)^3+x^3+(x+1)^3$$

has greater possibilities, for it can take an endless succession of values for varying values of x. When $x = 0, 1, 2, 3, 4$, for example, the values are 0, 9, 36, 99, 216. It is a 'living' expression, completely responsive to the changes which occur in the variable x.

An expression whose value depends in this way on a variable x is said to be a *function* of x.

The generality of a function, as compared with a fixed expression, often serves to reveal unsuspected properties common to groups of numbers. Thus, by direct multiplication,

$$\begin{aligned}&(x-1)^3+x^3+(x+1)^3\\ &\quad = (x^3-3x^2+3x-1)+x^3+(x^3+3x^2+3x+1)\\ &\quad = 3x(x^2+2).\end{aligned}$$

But the right-hand side has $3x$ as a factor. Hence *the sum of the cubes of three consecutive integers is exactly divisible by three times the middle integer.*

For example,

$$6^3+7^3+8^3 = 1071 = 21\times 51.$$

A function of a variable x may be conveniently denoted for reference by a single letter such as f; for instance,

$$f = (x-1)^3+x^3+(x+1)^3.$$

The dependence on x may be emphasized by the more extended notation

$$f(x),$$

read as 'f of x'. Other symbols, like

$$g(x),\quad h(x),\quad F(x),\quad U(x)$$

are also used.

The symbol $f(4)$ is used to denote *the value of* $f(x)$ *when* x *has the value* 4. Thus, if $f(x)$ is x^3-60, then

$$f(4) = (4)^3-60 = 4.$$

In the same way, when $f(x)$ is x^3-60,

$$\begin{aligned}f(-1) &= (-1)^3-60 = -61,\\ f(0) &= (0)^3-60 = -60,\\ f(3) &= (3)^3-60 = -33.\end{aligned}$$

Note, too, that, if $f(x)$ is x^3-60, then

$$\begin{aligned}f(x^2) &= (x^2)^3-60 = x^6-60,\\ f(\sin x) &= (\sin x)^3-60 = \sin^3 x-60.\end{aligned}$$

These and similar adaptations of the notation for a function will be incorporated without special reference.

3. Identities and inequalities; notation. The symbol $\equiv$ is used to denote *definition* or *identity*. Thus the statement

$$f(x) \equiv (x-1)^3+x^3+(x+1)^3$$

reads: *$f(x)$ is defined to be* $(x-1)^3+x^3+(x+1)^3$ for all values of x. Again, the statement

$$(x^2-1)^2+(2x)^2 \equiv (x^2+1)^2$$

reads: *the function* $(x^2-1)^2+(2x)^2$ *is identically equal to the function* $(x^2+1)^2$, *the relation being true for all values of* x; for example, the values $x = 2, 4, 6$ give the formulae ('Pythagoras' triangles)

$$3^2+4^2 = 5^2, \quad 15^2+8^2 = 17^2, \quad 35^2+12^2 = 37^2.$$

In both cases, the emphasis is on 'all values of x'.

Few writers are as correct as they ought to be in this matter, and this book is unlikely to be found entirely consistent. Strictly speaking, a statement

$$ax+b \equiv 3x+2$$

ought to mean that the two sides are equal for all values of x, so that $a = 3$, $b = 2$; and a statement

$$ax+b = 3x+2$$

ought to mean that there is a value of x such that

$$(a-3)x = 2-b,$$

the value being

$$x = \frac{2-b}{a-3}.$$

The first statement is usually given correctly; but in the second the sign of equality $=$ is sometimes used when the sign of identity or definition $\equiv$ is really meant. In practice, the context indicates which interpretation is intended.

The symbol $\neq$ is used to denote *inequality*. Thus the statement

$$x^2-3x+2 \neq 0$$

asserts that x must not have either of the values, 1, 2.

The symbol $>$ is used for *greater than*, and the symbol $\geqslant$ for *greater than or equal to*. Thus the statement

$$x^2 > 9 \quad \text{if} \quad x > 3$$

asserts that x^2 is greater than 9 whenever x is greater than 3; and the statement

$$x^2-2x+8 \equiv (x-1)^2+7 \geqslant 7$$

asserts that (since the square $(x-1)^2$ is always greater than or equal to zero) the function x^2-2x+8 is always greater than or equal to 7 in value.

The symbol $<$ is used for *less than*, and the symbol $\leqslant$ for *less than or equal to*.

A statement like $$1 \leqslant x < 5$$ means that x lies between 1 and 5, and further that x may be equal to 1.

One other symbol may be given at this point. The inclusion of a number between vertical lines $$|x|$$ means that the *numerical value* of x is to be taken.

Thus $$|-5| = 5, \quad |-\tfrac{1}{2}| = \tfrac{1}{2}, \quad |3| = 3.$$

With this usage, the inequality

$$-2 < x < 2$$

may be written $$|x| < 2,$$

and the inequality $$-3 \leqslant x \leqslant 3$$

may be written $$|x| \leqslant 3.$$

4. Zero. The 'number' zero has one characteristic property that must be clearly grasped: *if a product has value zero, then at least one of its factors must be zero.* This is the basis of the well-known routine used to complete the solution of a quadratic equation

$$(x-1)(x-2) = 0.$$

'Hence *either* $x-1 = 0$, *or* $x-2 = 0$;

that is, *either* $x = 1$, *or* $x = 2$.'

Observe, too, that division by zero is a meaningless operation; a symbol like

$$\tfrac{3}{0}$$

has no meaning. Again, *a relation*

$$a.0 = b$$

cannot hold unless $b = 0$, *in which case it is automatically true for all values of* a.

5. Polynomials. Let x be a variable, and $a, b, c, d, \ldots$ a succession of constants. The expressions

$$ax+b,$$
$$ax^2+bx+c,$$
$$ax^3+bx^2+cx+d,$$
$$\cdots\cdots\cdots\cdots\cdots\cdots$$

are called *polynomials* in x; a polynomial is formed as a sum of terms, each of which is a constant multiple of a power of x. The highest power of x is called the *degree* of the polynomial. Polynomials of degree 1, 2, 3, 4, 5, 6 are called linear, quadratic, cubic, quartic, quintic, sextic. The three polynomials at the top of this paragraph are linear, quadratic, cubic.

A polynomial is a simple example of a function of x. Its dependence on x may be exhibited to the eye by drawing the *graph* of the function. For instance, the polynomial x^2-2x+3 may be exhibited by the graph

$$y = x^2-2x+3$$

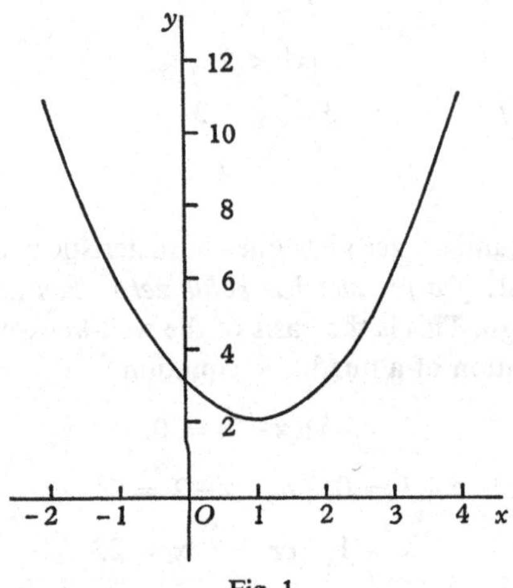

Fig. 1

shown in the diagram (fig. 1); the table, for values of x in integers between -2, 4, is

x	-2	-1	0	1	2	3	4
y	11	6	3	2	3	6	11

Some properties of the function

$$f(x) \equiv x^2-2x+3$$

may be deduced from the graph. For example,

(i) $f(x)$ is never less than 2 in value;

(ii) there are *two* values of x for which $f(x)$ takes any given value greater than 2;

(iii) since the graph is symmetrical about the line $x = 1$, the values of $f(x+1)$ and $f(-x+1)$ are equal for all values of x.

6. The zeros of polynomials; equations. Given a polynomial $f(x)$, particular importance attaches to those values (if any) of x for which

$$f(x) = 0.$$

The polynomial is then *equated to zero*, or, alternatively, *vanishes*. The values of x for which this happens are the *roots* of the equation.

An equation as first derived will probably not appear in a tidy form like

$$ax+b = 0,$$
$$ax^2+bx+c = 0.$$

Considerable algebraic manipulation may first be required to reduce it. But, once the preliminary manipulation is past, the equation (provided that it is of the polynomial type) must ultimately come to such a form; the method of solution then follows standard procedure, of which an account is given later.

***Illustration* 2.** *To solve the equation*

$$x-\tfrac{2}{3}(3x+4) = \tfrac{5}{6}-\tfrac{1}{2}(x+1).$$

Multiply throughout by 6:

$$6x-4(3x+4) = 5-3(x+1).$$

Remove brackets: $6x-12x-16 = 5-3x-3.$

Collect terms: $3x+18 = 0.$

Hence $x = -6.$

Examples 1

Reduce the following equations to linear or quadratic form, and then solve them.

1. $\tfrac{1}{2}(1-x) = \tfrac{1}{3}(1+x).$

2. $\dfrac{3}{1-x} = \dfrac{2}{1-3x}.$

3. $\dfrac{x-1}{2}+\dfrac{x-2}{3}+\dfrac{x-3}{4} = 0.$

4. $(x+3)(x-3) = 3x-11.$

5. $x(x-4) = -3.$

6. $\dfrac{x-\frac{1}{2}}{\frac{1}{2}}+\dfrac{x-\frac{2}{3}}{\frac{2}{3}} = 0.$

7. $\tfrac{1}{2}(x+2a) = \tfrac{1}{4}(x-2a).$

8. $\dfrac{x+a-b}{a} = \dfrac{x-a+b}{b}.$

9. $x(x-4) = x-6.$

10. $\dfrac{x}{3} = \dfrac{2}{7-x}.$

11. $\dfrac{x+3}{\frac{1}{2}}-\dfrac{x+1}{\frac{1}{3}} = \dfrac{x-1}{\frac{1}{4}}$

12. $x-\tfrac{1}{5}(3x-5) = 2(\tfrac{1}{6}-\tfrac{1}{3}x).$

13. $\tfrac{5}{3}(x-3)-2\{\tfrac{7}{2}(x-2)+3\} = 0.$

14. $x(x-1) = 6.$

7. Polynomials in several variables. The polynomials so far considered have all been functions of a single variable. The idea is easily extended. Expressions like

$$ax^2+bxy+cy^2+dx+ey+f,$$
$$x^3+y^3+x^2+y^2+x+y+1,$$
$$x^4y^2+3xy^3+5$$

are called *polynomials in* x, y; expressions like

$$x^3+y^2+z,$$
$$ax^2+by^2+cz^2+2fyz+2gzx+2hxy,$$
$$x^5y^2+y^4z^3+z^3x^4+14$$

are called *polynomials in* x, y, z. The characteristic feature of a polynomial is that it is a *sum of multiples of products of powers of the variables*; thus, for variables x, y, z, powers might be x^3, y, z^2, the product would be x^3yz^2, a multiple might be $5x^3yz^2$, and the polynomial is a sum of terms derived in this way.

The *order* of any term in the polynomial is the sum of the exponents of the powers of $x, y, z, \ldots$ in it; for example, the orders of terms like $3x^3yz^2$ and $7xy^5z^4$ are $(3+1+2) = 6$ and $(1+5+4) = 10$. The *degree* of a polynomial is the order of the term of highest order in it; for example, the three polynomials in x, y at the beginning of this paragraph have orders 2, 3, 6; and the polynomials in x, y, z have orders 3, 2, 7.

A polynomial is said to be *homogeneous* when all of its terms are of the same order. Typical examples are

$$2x+3y+7z,$$
$$5x^2+4xy+y^2-3xy,$$
$$x^3+y^3+z^3-3xyz.$$

8. The method of mathematical induction. This important method of investigation can be used for many purposes, and may be explained by two typical examples, of which the first is very elementary.

***Illustration* 3.** *To find a formula for the n-th odd number, in the sequence*

$$1, 3, 5, 7, 9, \ldots.$$

By inspection, the first, second, third odd numbers are

$$2.1-1, \quad 2.2-1, \quad 2.3-1$$

respectively, suggesting the formula

$$2n-1$$

for the nth. The method of induction proceeds by three stages:

(i) *Assume*, as a starting-point, that the formula $2n-1$ is valid up to some particular value k. Thus we *assume* that the kth odd number is

$$2k-1.$$

(ii) *Examine the* $(k+1)$*th term on the basis of this assumption*. The $(k+1)$th odd number, by definition, exceeds the kth by 2, so that it is

$$2k+1,$$

or

$$2(k+1)-1.$$

(iii) *Observe* that the expression for the $(k+1)$th term is precisely the same as that for the kth with the number k replaced by $k+1$. Hence if the formula $2n-1$ is true, in the form $2k-1$, for $n = k$, it is necessarily true, in the form $2(k+1)-1$, for $n = k+1$.

But it is true, in the form $2.1-1$, for $n = 1$. Hence it is true for $n = 2$; hence it is true for $n = 3$; for $n = 4$; ...; and, successively, for all values of n. The formula $2n-1$ is therefore established.

Illustration **4.** *To find a formula for the sum*

$$1+3+5+7+...+(2n-1)$$

of the first n odd numbers.

Write S_1, S_2, S_3, ... to denote the sums of the first 1, 2, 3, ... odd numbers. Then

$$S_1 = 1,$$
$$S_2 = 1+3 = 4 = 2^2,$$
$$S_3 = 1+3+5 = 9 = 3^2,$$
$$S_4 = 1+3+5+7 = 16 = 4^2,$$

suggesting the formula

$$S_n = n^2.$$

(i) Assume, as a starting-point, that the formula $S_n = n^2$ is valid up to some particular value k. Thus we *assume* that

$$S_k = k^2.$$

(ii) Examine S_{k+1} on the basis of this assumption. The sum S_{k+1} is found by adding to S_k the $(k+1)$th odd integer, so that (as above)

$$S_{k+1} = k^2+(2k+1)$$
$$= (k+1)^2.$$

(iii) The expression for S_{k+1} is precisely that for S_k with k replaced by $k+1$. Hence if the formula is true for $n = k$, it is necessarily true for $k+1$.

But it is true for $n = 1$, since $S_1 = 1^2$. Hence it is true for S_2; hence it is true for S_3; for S_4; ...; and, successively, for all values of n.

The formula $S_n = n^2$ is therefore established.

Examples 2

Use the method of mathematical induction to establish the following results:

1. The sum of the first n integers is $\frac{1}{2}n(n+1)$.

2. The sum of the first n powers of 2,

$$1+2+4+8+\ldots+2^{n-1},$$

is 2^n-1.

3. The sum of the first n even integers

$$2+4+6+8+\ldots+2n$$

is $n(n+1)$.

4. The number of straight lines joining all possible pairs of n points (in general position) is $\frac{1}{2}n(n-1)$.

[For example, the points A, B, C, D are joined in pairs by the six lines BC, CA, AB, AD, BD, CD.]

5. The number $3.5^{2n+1}+2^{3n+1}$ is divisible by 17.

6. The number $3^{2n+2}-8n-9$ is divisible by 64.

7. The number $3n^5+7n$ is a multiple of 5; hence $n^2(n^2+1)(n^2+4)$ is also a multiple of 5.

8. The sum

$$1^5+2^5+3^5+\ldots+n^5$$

is

$$\tfrac{1}{12}n^2(n+1)^2(2n^2+2n-1).$$

2

THE LINEAR POLYNOMIAL

The properties recorded here are given, not so much for their intrinsic worth, as because they are the first cases in inductions which extend to polynomials of higher degree and greater complexity.

1. Polynomial with a given zero. To prove that *a linear polynomial which vanishes for $x = p$ can necessarily be expressed in the form*

$$a(x-p),$$

where a is a constant.

The linear polynomial is, by definition,

$$f(x) \equiv ax+b.$$

Since $f(p) = 0$, given, it follows that

$$ap+b = 0,$$

or

$$b = -ap.$$

Hence

$$f(x) \equiv ax-ap$$

$$\equiv a(x-p).$$

The value of the constant a, if required, must be determined by some other condition. For example, if the polynomial has value M when $x = q$ (with $q \neq p$), then

$$a(q-p) = M,$$

so that

$$a = \frac{M}{q-p},$$

and the polynomial is

$$M\left(\frac{x-p}{q-p}\right).$$

2. Linear polynomial with two (distinct) zeros. To prove that *a linear polynomial which vanishes for two values $x = p$, $x = q$ (with $p \neq q$) is identically zero.*

By identically zero we mean (compare p. 4) that it is zero *for every value of x.* For example, the polynomial $(a-2)x-(a^2-4)$ happens to have the value zero in the particular case when $x = a+2$; exceptionally, however, a may have the value 2, in which case the polynomial is *identically* zero, being zero for all values of x.

Since it vanishes for $x = p$, the linear polynomial, by § 1, is expressible in the form

$$f(x) \equiv a(x-p),$$

where a is a constant. Since $f(q)$ also vanishes,

$$a(q-p) = 0.$$

Since $q \neq p$, the factor $q-p$ is not zero. Hence (p. 5)

$$a = 0,$$

and so the polynomial is identically zero.

3. Unique determination. To prove that *a linear polynomial is determined, and determined uniquely, when it takes assigned values for two distinct values of x.*

Suppose that the polynomial has values L, M for $x = p, q$, with $p \neq q$. The polynomial may be written down with the help of a device which should be remembered carefully for use in more elaborate cases. Observe that

(i) the function $$\frac{x-q}{p-q}$$
has values 1, 0 for $x = p, q$,

(ii) the function $$\frac{x-p}{q-p}$$
has values 0, 1 for $x = p, q$.

Thus the expression

$$L\left(\frac{x-q}{p-q}\right)+M\left(\frac{x-p}{q-p}\right),$$

which is, on expansion, the linear polynomial

$$\left(\frac{L-M}{p-q}\right)x+\left(\frac{Mp-Lq}{p-q}\right),$$

takes the required values L, M for $x = p, q$.

To prove that this expression is unique, assume that, on the contrary, there are two such polynomials, $ax+b$ and $a'x+b'$.

Then
$$ap+b = L = a'p+b',$$
so that
$$(a-a')p+(b-b') = 0;$$
and
$$aq+b = M = a'q+b',$$
so that
$$(a-a')q+(b-b') = 0.$$

Hence the linear polynomial

$$(a-a')x+(b-b')$$

is zero for $x = p$ and for $x = q$ (with $p \neq q$), so that (§2) it is identically zero. Hence

$$a = a', \quad b = b'$$

and so the two proposed polynomials are in fact the same.

***Illustration* 1.** *To write down the linear polynomial which takes the values* $-6, 5$ *when* $x = 1, -2$. The polynomial is

$$-6\left(\frac{x+2}{1+2}\right)+5\left(\frac{x-1}{-2-1}\right),$$

or
$$-\tfrac{6}{3}(x+2)-\tfrac{5}{3}(x-1),$$

or
$$-\tfrac{11}{3}x-\tfrac{7}{3}.$$

4. The linear equation. The solution of the linear equation

$$ax+b = 0$$

is almost obvious, namely $\quad x = -\dfrac{b}{a}.$

There is one point of difficulty, which ought not to be emphasized at present but which is nevertheless worth recording as it is basically important in more advanced work. The equation

$$ax+b = 0$$

can always be solved, as stated, *except when a is zero* (cf. p. 5). If

$$a = 0, \quad b \neq 0,$$

the equation
$$0.x+b = 0$$

cannot be satisfied by any value of x; and if $a = 0, b = 0$, the equation

$$0.x+0 = 0$$

is satisfied for *every* value of x.

***Illustration* 2.** The case of exception may seem trivial; perhaps it is. But the approach to it may come along lines which give it deeper significance:

To solve the equation
$$(p^2-3p+2)x = q-3.$$

The solution is
$$x = \frac{q-3}{p^2-3p+2}$$

except for any values of p which make $p^2-3p+2 = 0$; and two such exceptional values exist, namely $p = 1, p = 2$.

If $p = 1$, or if $p = 2$, the equation has no solution in x, unless also $q = 3$, in which case it is satisfied for every value of x. The solutions may be exhibited by the table:

Value of p	Value of q	Solution
Not 1 or 2	Anything	$\dfrac{q-3}{p^2-3p+2}$
1 or 2	Anything except 3	No solution
1 or 2	3	x can have any value

Examples 1

Find linear polynomials subject to the following conditions:

1. Value 2 for $x = 1$, value 1 for $x = 2$.
2. Value -3 for $x = 2$, value $+3$ for $x = -2$.
3. Value $\frac{1}{2}$ for $x = 4$, value 2 for $x = \frac{1}{4}$.
4. Value 0 for $x = 1$, value 1 for $x = 0$.
5. Value 3 for $x = -2$, value 3 for $x = 5$.
6. Value $\frac{1}{3}$ for $x = 3$, value 0 for $x = \frac{1}{3}$.
7. Value a for $x = b$, value b for $x = a$.
8. Value $2a$ for $x = a$, value $3a$ for $x = 2a$.
9. Value 0 for $x = 1$, value 0 for $x = 2$.
10. Value a^2 for $x = a$, value b^2 for $x = b$.
11. Value -3 for $x = -3$, value -4 for $x = -4$.
12. Value 1 for $x = a$, value 2 for $x = b$.

Solve completely, after the manner of Illustration 2, the equations:

13. $px = 1$.
14. $(p-1)(p-2)x = 1$.
15. $px = p+1$.
16. $(p-2)x = q$.
17. $p^2x = q+3$.
18. $p(p+4)x = q+5$.
19. $(a^2-b^2)x = 1$.
20. $(a-2)(a-4)x = b+6$.

3

THE QUADRATIC POLYNOMIAL

The results of chapter 2 may be extended to quadratic polynomials.

1. The quadratic polynomial with two given zeros. To prove that *a quadratic polynomial which vanishes for $x = p$, $x = q$, with $p \neq q$, can necessarily be expressed in the form*

$$a(x-p)(x-q).$$

[The converse result, that this polynomial does vanish for $x = p$, $x = q$, is obvious.]

Suppose that such a quadratic is

$$ax^2+bx+c,$$

and form the function

$$ax^2+bx+c-a(x-p)(x-q).$$

By definition, $ap^2+bp+c = 0, \quad aq^2+bq+c = 0,$

so that the function vanishes for $x = p$, $x = q$. But the function is actually, by expansion, the *linear* polynomial

$$\{b+a(p+q)\}x+(c-apq) = 0,$$

and, since it vanishes for the two values $x = p$, $x = q$, it is (p. 11) identically zero. Hence

$$ax^2+bx+c-a(x-p)(x-q) \equiv 0,$$

so that $\quad ax^2+bx+c \equiv a(x-p)(x-q),$

giving the required form.

The value of a, if required, must be determined by other conditions.

2. The quadratic polynomial with three (distinct) zeros. To prove that *a quadratic polynomial which vanishes for three values $x = p$, $x = q$, $x = r$ (with p, q, r unequal) is necessarily zero.*

By § 1, the polynomial is of the form

$$a(x-p)(x-q).$$

It vanishes for $x = r$; hence

$$a(r-p)(r-q) = 0.$$

But $r-p \neq 0$, $r-q \neq 0$, so that (p. 5)

$$a = 0.$$

Hence the polynomial is identically zero.

3. Unique determination. To prove that *a quadratic polynomial can be determined, and determined uniquely, so as to have assigned values for three distinct values of x.*

Suppose that the polynomial has values L, M, N for $x = p, q, r$ (with p, q, r unequal). It can be written down, after the manner illustrated for the linear polynomial (p. 12), in the form

$$L\left(\frac{x-q}{p-q}\right)\left(\frac{x-r}{p-r}\right)+M\left(\frac{x-r}{q-r}\right)\left(\frac{x-p}{q-p}\right)+N\left(\frac{x-p}{r-p}\right)\left(\frac{x-q}{r-q}\right).$$

This is verified at once by putting $x = p, q, r$ in turn.

Moreover, it is *unique*. For suppose that there were two such polynomials, say

$$ax^2+bx+c, \quad a'x^2+b'x+c'.$$

Then

$$ap^2+bp+c = L = a'p^2+b'p+c',$$

so that

$$(a-a')p^2+(b-b')p+(c-c') = 0,$$

with similar results for q, r. Hence the polynomial

$$(a-a')x^2+(b-b')x+(c-c')$$

is zero for the three values $x = p, q, r$, and so, by §2, is identically zero. Hence

$$a = a', \quad b = b', \quad c = c'.$$

Thus the two proposed polynomials are in fact the same.

Illustration 1. *To write down the quadratic expression which takes the values* 12, 2, 24 *when* $x = 2, 0, -2$.

The expression is

$$12\frac{(x-0)(x+2)}{(2-0)(2+2)}+2\frac{(x+2)(x-2)}{(0+2)(0-2)}+24\frac{(x-2)(x-0)}{(-2-2)(-2-0)}$$

$$= \tfrac{3}{2}(x^2+2x)-\tfrac{1}{2}(x^2-4)+3(x^2-2x)$$

$$= 4x^2-3x+2,$$

after simplification.

Examples 1

Find the quadratic polynomials determined by the following conditions.

1. Values 6, 2, 6 for $x = -1, 3, 4$.
2. Values 1, 3, 1 for $x = -1, 1, 0$.
3. Values 6, 4, 9 for $x = -2, 0, 1$.
4. Values 0, -2, -2 for $x = 1, 2, 3$.
5. Values 0, 0, 1 for $x = -3, -2, 0$.
6. Values 0, -1, 20 for $x = 1, 0, 3$.

It is assumed that the reader can factorize quadratic polynomials with reasonably fluency in suitable cases, and hence solve simple quadratic equations. The method which follows, leading to a general formula, is probably familiar, but is inserted for completeness.

4. The general solution of the quadratic equation. Consider the equation

$$ax^2+bx+c = 0.$$

Divide by a, and transfer the constant term to the right-hand side:

$$x^2+\left(\frac{b}{a}\right)x = -\left(\frac{c}{a}\right).$$

Add $\left(\frac{b}{2a}\right)^2$—the square of half the (new) coefficient of x—to either side:

$$x^2+\left(\frac{b}{a}\right)x+\left(\frac{b}{2a}\right)^2 = \left(\frac{b}{2a}\right)^2-\left(\frac{c}{a}\right) = \frac{b^2-4ac}{4a^2}.$$

Observe that the left-hand side is $\left(x+\frac{b}{2a}\right)^2$. Thus

$$\left(x+\frac{b}{2a}\right)^2 = \frac{b^2-4ac}{4a^2}.$$

Remember that, if two numbers are equal, then their square roots are either equal or equal in value but opposite in sign. Thus

$$\text{either} \quad x+\frac{b}{2a} = +\frac{\surd(b^2-4ac)}{2a},$$

$$\text{or} \quad x+\frac{b}{2a} = -\frac{\surd(b^2-4ac)}{2a}.$$

The double choice is often denoted in a single equation by the notation

$$x+\frac{b}{2a} = \pm\frac{\surd(b^2-4ac)}{2a},$$

and it follows at once that

$$x = \frac{-b \pm \sqrt{(b^2-4ac)}}{2a}.$$

Hence *the two roots of the equation*

$$ax^2+bx+c = 0$$

are

$$\frac{-b \pm \sqrt{(b^2-4ac)}}{2a}.$$

***Illustration* 2.** *To solve the equation*

$$2x^2+3x-4 = 0.$$

Divide by 2 and transfer the constant:

$$x^2+\tfrac{3}{2}x = 2.$$

Complete the square on the left-hand side by adding $(\frac{3}{4})^2$ to each side:

$$x^2+\tfrac{3}{2}x+(\tfrac{3}{4})^2 = 2+(\tfrac{3}{4})^2 = 2\tfrac{9}{16},$$

or

$$(x+\tfrac{3}{4})^2 = \tfrac{41}{16}.$$

Take square roots:

$$x+\tfrac{3}{4} = \frac{\pm\sqrt{41}}{4},$$

so that *either*

$$x = \frac{-3+\sqrt{41}}{4},$$

or

$$x = \frac{-3-\sqrt{41}}{4}.$$

Examples 2

Solve, to 2 places of decimals, the following equations:

1. $x^2-2x-4 = 0$.
2. $2x^2-3x-4 = 0$.
3. $x^2+10x+1 = 0$.
4. $x^2-8x+2 = 0$.
5. $4x^2-9x-3 = 0$.
6. $2x^2-7x+4 = 0$.
7. $3x^2-3x-11 = 0$.
8. $5x^2-12x+2 = 0$.

5. Equal roots. The two roots

$$\frac{-b+\sqrt{(b^2-4ac)}}{2a}, \quad \frac{-b-\sqrt{(b^2-4ac)}}{2a}$$

of the equation

$$ax^2+bx+c = 0$$

are distinct, except when

$$b^2-4ac = 0,$$

in which case they have the common value $(-b/2a)$. It is customary to say that the quadratic equation then has a *repeated root*, or *equal roots*.

The earlier results of this chapter concerned values of x necessarily unequal. The result of §1, however, can be proved true in the modified form:

Every quadratic polynomial which is zero when $x = p$, and only when $x = p$, can be written in the form

$$a(x-p)^2.$$

If the quadratic is $$ax^2+bx+c,$$

then $$b^2-4ac = 0,$$

since, as above, there would otherwise be two distinct roots; and the common value is $-b/2a$, so that, here,

$$b = -2ap.$$

Thus $$4ac = b^2 = 4a^2p^2.$$

Now a cannot be zero, or the polynomial would not be quadratic; hence

$$c = ap^2.$$

The quadratic is therefore

$$ax^2-2apx+ap^2,$$

or $$a(x-p)^2.$$

Examples 3

Find the value or values of k for the following quadratic equations to have equal roots.

1. $x^2+4x+k = 0.$
2. $kx^2-2x+1 = 0.$
3. $kx^2+9x+k = 0.$
4. $x^2+kx+36 = 0.$
5. $x^2+kx+144 = 0.$
6. $4x^2-kx+9 = 0.$
7. $kx^2+6x+9 = 0.$
8. $x^2+22x+k = 0.$
9. $k(x^2-10x-2)+(2x^2+1) = 0.$

6. Complex roots. One point of difficulty in the solution of quadratic equations leads to a very wide piece of theory (*complex numbers*) which we do little more than mention here. Details will be given in chapter 4.

The immediate problem arises in connection with an equation of the type

$$x^2+2x+5 = 0.$$

The process of §4 is $$x^2+2x = -5,$$
or $$x^2+2x+1 = -4,$$
or $$(x+1)^2 = -4.$$

Now comes a stop, for no number exists whose square is -4; the equation has *no solutions.*

More generally, the solutions

$$\frac{-b+\sqrt{(b^2-4ac)}}{2a}, \quad \frac{-b-\sqrt{(b^2-4ac)}}{2a}$$

do not exist when b^2-4ac is negative, for the square roots cannot then be taken.

It will therefore be implicit in what follows that *the coefficients in the equation*

$$ax^2+bx+c = 0$$

are so chosen that the number b^2-4ac is not negative.

The work of chapter 4 will be aimed at devising an extended system of numbers by means of which solutions of this equation can be obtained. In terms of these numbers, the equation will be said to have *complex roots*; but the present paragraph is merely a warning in case the reader should meet them prematurely.

7. The coefficients of a quadratic equation in terms of the roots. Suppose that the roots of the quadratic equation

$$ax^2+bx+c = 0 \quad (a \neq 0)$$

are p, q. Then the equation is (p. 15) the same as

$$a(x-p)(x-q) = 0,$$
or $$ax^2-a(p+q)x+apq = 0,$$
so that $$b = -a(p+q),$$
$$c = apq.$$

Hence *the sum and the product of the roots of the equation $ax^2+bx+c = 0$ are given by the formulae*

$$p+q = -b/a,$$
$$pq = c/a.$$

Conversely, *the equation whose roots are p, q may be written down by the formula*

$$x^2-x\ (\text{sum of roots})+(\text{product of roots}) = 0.$$

The following illustrations give typical examples of the use of these results. Alternative methods for similar problems are also given in chapter 5.

Illustration 3. *To form the equation whose roots are* 3 *times those of the equation*

$$x^2-5x-7 = 0.$$

If p, q are the roots of the given equation, then

$$p+q = 5, \quad pq = -7.$$

If p', q' are the roots of the required equation, then

$$p'+q' = 3p+3q = 15, \quad p'q' = 9pq = -63.$$

Hence the required equation is

$$x^2-15x-63 = 0.$$

Illustration 4. *To form the equation whose roots are* $p+q^2$, p^2+q, *where* p, q *are the roots of the equation*

$$x^2+mx+n = 0.$$

Write $$p' \equiv p+q^2, \quad q' \equiv p^2+q.$$

Then $$p'+q' = (p+q)+(p^2+q^2)$$
$$= (p+q)+(p+q)^2-2pq,$$

and $$p'q' = (p+q^2)(p^2+q) = p^3+q^3+pq+p^2q^2$$
$$= (p+q)\{(p+q)^2-3pq\}+pq+p^2q^2.$$

Now $$p+q = -m, \quad pq = n,$$

so that $$p'+q' = -m+m^2-2n,$$
$$p'q' = -m(m^2-3n)+n+n^2.$$

Hence the required equation is

$$x^2+(m-m^2+2n)x-m^3+3mn+n+n^2 = 0.$$

Examples 4

1. Form the equation whose roots are twice those of $2x^2+7x-9 = 0$.

2. Form the equation whose roots exceed by 3 those of

$$3x^2+8x+3 = 0.$$

3. Form the equation whose roots are the squares of those of $5x^2-x-7 = 0$.

4. Form the equation whose roots are 5 less than those of

$$x^2+10x+1 = 0.$$

5. The roots of the equation $x^2+mx+n = 0$ are p, q. Form the equations whose roots are:

(i) $p+2, q+2$.
(ii) $pq+p, pq+q$.
(iii) p^2, q^2.
(iv) $1/p, 1/q$.
(v) $p+1/p, q+1/q$.
(vi) p^2-1, q^2-1.
(vii) p^3, q^3.
(viii) $1/p^2, 1/q^2$.
(ix) $p+2q, 2p+q$.
(x) $p^2+2q^2, 2p^2+q^2$.

6. Repeat example 5 when p, q are the roots of the equation $ax^2+bx+c = 0$.

4

COMPLEX NUMBERS

1. Introduction. It may be helpful if we begin this chapter by reminding the reader of the types of elements which he is already accustomed to use in arithmetic. These are

(i) the *integer* (positive or negative);

(ii) the *rational number*, that is, the ratio of two integers;

(iii) the *irrational number* (such as $\sqrt{2}, \pi$) which cannot be expressed as a ratio of integers.

It is now necessary to extend our scope and to devise a system of numbers not included in any of these three classes.

In order to exhibit the need for the new numbers, we solve in succession a series of quadratic equations, similar to look at, but essentially distinct.

I. $$x^2-6x+5 = 0.$$

Following the usual 'completing the square' process, we have the solution

$$\begin{aligned} x^2-6x &= -5, \\ x^2-6x+9 &= -5+9, \\ &= 4, \\ (x-3)^2 &= 4. \end{aligned}$$

The important point at this stage is the existence of two integers, namely $+2$ and -2, whose square is equal to 4. Hence we can find *integral* values of x to satisfy the equation:

$$x-3 = +2 \quad \text{or} \quad x-3 = -2,$$

so that $$x = 5 \quad \text{or} \quad x = 1.$$

The equation $x^2-6x+5 = 0$ can therefore be solved by taking x as one or other of the *rational* numbers (integers, in fact) 5 or 1.

II. $$x^2-6x+7 = 0.$$

Proceeding as before, we have the solution

$$\begin{aligned} x^2-6x &= -7, \\ x^2-6x+9 &= -7+9, \\ &= 2, \\ (x-3)^2 &= 2. \end{aligned}$$

Now comes a break; for there is no rational number whose square is 2, and so we cannot find a rational number x to satisfy the given equation. We must therefore extend our idea of number beyond the elementary realm of integers and rational numbers. This is an advance which the reader absorbed, doubtless unconsciously, many years ago, but it represents a step of fundamental importance. The theory of the irrational numbers will be found in text-books of analysis; for our purpose, knowledge of its existence is sufficient. In particular, we regard as familiar the concept of the irrational number, written as $\sqrt{2}$, whose value is 1·414..., with the property that $(\sqrt{2})^2 = 2$.

Once the irrational numbers are admitted, the solution of the equation follows:

$$x-3 = +\sqrt{2} \quad \text{or} \quad x-3 = -\sqrt{2},$$

so that

$$x = 3+\sqrt{2} \quad \text{or} \quad x = 3-\sqrt{2}.$$

The equation $x^2-6x+7 = 0$ can therefore be solved by taking x as one or other of the *irrational* numbers $3+\sqrt{2}$ or $3-\sqrt{2}$.

III.

$$x^2-6x+10 = 0.$$

As before,

$$\begin{aligned} x^2-6x &= -10, \\ x^2-6x+9 &= -10+9, \\ &= -1, \\ (x-3)^2 &= -1. \end{aligned}$$

Again comes the break; for there is no number, rational or irrational, whose square is the negative number -1. We have therefore two choices, to accept defeat and say that the equation has no solution, or to invent a new set of numbers from which a value of x can be selected. The second alternative is the purpose of this chapter.

To be strictly logical, we should now proceed to define these new numbers and then show how to frame the solution from among them. But their definition is, naturally, somewhat complicated, and it seems better to begin by merely postulating their 'existence'; we can then see what properties they will have to possess, and the reasons for the definition will become more apparent.

Just as, in earlier days, we learned to write the symbol '$\sqrt{2}$' for a number whose square is 2, so now we use the symbol '$\sqrt{(-1)}$' for a 'number' (in some sense of the word) whose square is -1; but, in practice, it is more convenient to have a single mark instead of the five marks $\sqrt{}$, (, $-$, 1,), and so we write

$$i \equiv \sqrt{(-1)}$$

as a convenient abbreviation.

We shall subject this symbol i to all the usual laws of algebra, but first endow it further with the property that

$$i^2 = -1.$$

This will be its sole distinguishing feature—though, of course, that feature is itself so overwhelming as to introduce us into an entirely new number-world.

It may be noticed at once that the 'number' $-i$ also has -1 for its square, since, in accordance with the normal rules,

$$\begin{aligned}(-i)^2 &= (-1\times i)^2 = (-1)^2\times(i)^2 = (+1)\times(-1)\\ &= -1.\end{aligned}$$

We round off this introduction by displaying the solution of the above equation:

$$x-3 = i \quad \text{or} \quad x-3 = -i,$$

so that

$$x = 3+i \quad \text{or} \quad x = 3-i.$$

By selecting x from this extended range of 'numbers', we are therefore able to find two solutions of the equation.

It may, perhaps, make this statement clearer if we verify how $3+i$, say, does exactly suffice. By saying that $3+i$ is a solution of the equation $x^2-6x+10 = 0$, we mean that

$$(3+i)^2-6(3+i)+10 = 0.$$

The left-hand side is

$$\begin{aligned}&9+6i+i^2-18-6i+10\\ &\qquad = 1+i^2\\ &\qquad = 1+(-1)\\ &\qquad = 0\\ &\qquad = \text{right-hand side,}\end{aligned}$$

so that the solution is verified.

We now investigate the elementary properties of our extended number-system, and then, when the ideas are a little more familiar, return to put the basis on a surer foundation.

Examples 1

Solve after the manner of the text the following sets of equations:

1. $x^2-4x+3 = 0,\quad x^2-4x+1 = 0,\quad x^2-4x+5 = 0.$

2. $x^2+8x+15 = 0,\quad x^2+8x+11 = 0,\quad x^2+8x+20 = 0.$

3. $x^2-2x-3 = 0,\quad x^2-2x-4 = 0,\quad x^2-2x+10 = 0.$

2. Definitions. The numbers of ordinary arithmetic (positive or negative integers, rational numbers and irrational numbers) are called *real* numbers.

If a, b are two real numbers, then numbers of the form

$$a+ib \quad (i^2 = -1)$$

are called *complex* numbers. When $b = 0$, such a number is *real*. When $a = 0$, the number

$$ib \quad (b \text{ real})$$

is called a *pure imaginary* number.

Two numbers such as

$$a+ib, \quad a-ib \quad (a, b \text{ real})$$

are called *conjugate* complex numbers.

A single symbol, such as c, is often used for a complex number. If

$$c \equiv a+ib,$$

then a is called the *real part* of c and b the *imaginary part*. The notations

$$a = \mathscr{R}c, \quad b = \mathscr{I}c$$

are often used.

The conjugate of a complex number c is often denoted by $\bar{c}$, so that, if $c = a+ib$, then $\bar{c} = a-ib$. Clearly the conjugate of $\bar{c}$ is c itself.

If c is *real*, then $c = \bar{c}$.

The magnitude

$$+\sqrt{(a^2+b^2)} \quad (a, b \text{ real})$$

is called the *modulus* of the complex number $c \equiv a+ib$; it is often denoted by the notations

$$|c|, \quad |a+ib|.$$

It follows that

$$|\bar{c}| = |c|.$$

Moreover, we have the relation

$$c\bar{c} = |c|^2,$$

for

$$\begin{aligned} c\bar{c} &= (a+ib)(a-ib) \\ &= a^2 - i^2b^2 = a^2-(-1)b^2 \\ &= a^2+b^2. \end{aligned}$$

Finally, since

$$c = a+ib, \quad \bar{c} = a-ib,$$

we have the relations

$$\mathscr{R}c = a = \tfrac{1}{2}(c+\bar{c}),$$

$$\mathscr{I}c = b = \frac{1}{2i}(c-\bar{c}).$$

Note. All the numbers now to be considered are actually complex, of the form $a+ib$, even when $b = 0$. Correctly speaking we should use the phrase 'real complex number' for a number such as 2, and 'pure imaginary complex number' for $2i$. But this becomes tedious once it has been firmly grasped that *all* numbers are complex anyway.

Examples 2

The following examples are designed to make the reader familiar with the use of the symbol i. Use normal algebra, except that $i^2 = -1$.

Express the following complex numbers in the form $a+ib$, where a, b are both real:

1. $(3+5i)+(7-2i)$. **2.** $(-2+3i)-(6-5i)$.

3. $(4+5i)(6-2i)$. **4.** $(2+7i)(-1+2i)$.

5. $3+2i+i(5+i)$. **6.** $4-3i+i(3+4i)$.

7. $(1+i)^2$. **8.** $(2+i)^3$.

9. $(2-i)^2-(3+2i)^2$. **10.** $(1+i)(1+2i)(1+3i)$.

Prove the following identities:

11. $\dfrac{1}{i} = -i$. **12.** $\dfrac{1}{1+i} = \tfrac{1}{2}(1-i)$.

13. $\dfrac{1}{3-4i} = \tfrac{1}{25}(3+4i)$. **14.** $\dfrac{1}{a+ib} = \dfrac{a-ib}{a^2+b^2}$.

15. $\dfrac{p+iq}{a+ib} = \dfrac{(ap+bq)+i(aq-bp)}{a^2+b^2}$.

3. Addition, subtraction, multiplication.

Let

$$c \equiv a+ib, \quad w \equiv u+iv \quad (a, b, u, v \text{ real})$$

be two given complex numbers. In virtue of the meaning which we have given to i, we obtain expressions for their sum, difference and product as follows:

(i) *Sum* $\quad c+w = (a+u)+i(b+v)$.

(ii) *Difference* $\quad c-w = (a-u)+i(b-v)$.

(iii) *Product*

$$\begin{aligned} cw &= (a+ib)(u+iv) \\ &= au+ibu+iav+i^2bv \\ &= (au-bv)+i(bu+av). \end{aligned}$$

The awkward one of these is the product. At present it is best not to remember the formula, but to be able to derive it when required.

Examples 3

Express the following products in the form $a+ib$ (a, b real):

1. $(2+3i)(4+5i)$.
2. $(1+4i)(2-6i)$.
3. $(-1-i)(-3+4i)$.
4. $(a+ib)(a-ib)$.
5. $i(4+3i)$.
6. $(\cos A+i\sin A)(\cos B+i\sin B)$.
7. $(3-i)(3+i)$.
8. $(2+3i)^3$.

4. Division. As before, let

$$c \equiv a+ib, \quad w \equiv u+iv \quad (a, b, u, v \text{ real}).$$

Then the quotient is, by definition, the fraction

$$\frac{c}{w} \equiv \frac{a+ib}{u+iv}.$$

It is customary to express such fractions in a form in which the denominator is real. For this, multiply numerator and demoninator by $\bar{w} \equiv u-iv$. Then

$$\begin{aligned}\frac{c}{w} &= \frac{(a+ib)(u-iv)}{(u+iv)(u-iv)}\\ &= \frac{(au+bv)+i(bu-av)}{u^2+v^2}\\ &= \frac{au+bv}{u^2+v^2}+i\frac{bu-av}{u^2+v^2}.\end{aligned}$$

It is implicit that u, v are not both zero.

For example, $(2+i)\div(3-i)$

$$\begin{aligned}&= \frac{2+i}{3-i}\\ &= \frac{(2+i)(3+i)}{(3-i)(3+i)}\\ &= \frac{6+5i+i^2}{9-i^2} = \frac{6+5i-1}{9+1}\\ &= \frac{5+5i}{10}\\ &= \tfrac{1}{2}+\tfrac{1}{2}i.\end{aligned}$$

Examples 4

Express the following quotients in the form $a+ib$ (a, b real):

1. $\dfrac{1+i}{1-i}$. 2. $\dfrac{3+4i}{4+3i}$. 3. $\dfrac{5i+6}{7i}$.

4. $\dfrac{2-3i}{4+i}$. 5. $\dfrac{3+5i}{1-6i}$. 6. $\dfrac{\cos 2\theta+i\sin 2\theta}{\cos\theta+i\sin\theta}$.

5. Equal complex numbers. To verify that, *if two complex numbers are equal, then their real parts are equal and their imaginary parts are equal.*

Let

$$c \equiv a+ib, \quad r \equiv p+iq$$

be two given complex numbers (a, b, p, q all real), such that $c = r$. Then

$$a+ib = p+iq,$$

or

$$a-p = i(q-b).$$

Squaring each side, we have

$$\begin{aligned}(a-p)^2 &= i^2(q-b)^2 \\ &= -(q-b)^2.\end{aligned}$$

But $a-p$, $q-b$ are real, so that their squares are positive or zero; hence this relation cannot hold unless each side is zero. That is,

$$a = p, \quad b = q,$$

as required.

COROLLARY. *If a, b are real and*

$$a+ib = 0,$$

then

$$a = 0, \quad b = 0.$$

Examples 5

Find the sum, difference, product and quotient of each of the following pairs of complex numbers:

1. $2+3i$, $3-5i$. 2. $-4+2i$, $-3+7i$.

3. 4, $2i$. 4. $3+2i$, $-i$.

5. $2+3i$, $2-3i$. 6. $-3+4i$, $-3-4i$.

Find, in the form $a+ib$ (a, b real), the reciprocal of each of the following complex numbers:

7. $3+4i$. 8. $-5+12i$.

9. $6i$. 10. $-6-8i$.

Solve the following quadratic equations, expressing your answers in the form $a \pm ib$:

11. $x^2+4x+13 = 0$. **12.** $x^2-2x+2 = 0$.

13. $x^2+6x+10 = 0$. **14.** $4x^2+9 = 0$.

15. $x^2-8x+25 = 0$. **16.** $x^2+4x+5 = 0$.

6. The complex number as a number-pair. A complex number

$$c \equiv a+ib$$

consists essentially of the pair of real numbers a, b linked by the symbol i to which we have given a specific property (i^2 negative) not enjoyed by any real number. Hence c is of the nature of an *ordered pair* of real numbers, the ordering being an important feature since, for example, the two complex numbers $4+5i$ and $5+4i$ are quite distinct.

The concept of number-pair forms the foundation for the more logical development promised at the end of §1, for it is precisely this concept which enables us to extend the range of numbers required for the solution of the third quadratic equation ($x^2-6x+10 = 0$). We therefore make a fresh start, with a series of definitions designed to exhibit the properties hitherto obtained by the use of the fictitious 'number' i.

We define a *complex number* to be an ordered number-pair (of *real* numbers), denoted by the symbol $[a, b]$, subject to the following rules of operation (chosen, of course, to fit in to the treatment given at the start of this chapter):

(i) The symbols $[a, b]$, $[u, v]$ are equal if, and only if, the two relations $a = u$, $b = v$ are satisfied;

(ii) The sum

$$[a, b]+[u, v]$$

is the number-pair $[a+u, b+v]$;

(iii) The product

$$[a, b]\times[u, v]$$

is the number-pair $[au-bv, bu+av]$.

Compare the formulae in §3.

A number $[a, 0]$, whose second component is zero, is called a *real complex number*; a number $[0, b]$, whose first component is zero, is called a *pure imaginary complex number*. These terms are usually abbreviated to *real* and *pure imaginary* numbers. (Compare p. 26.)

In order to achieve correlation with the normal notation for real numbers, and with the customary notation based on the letter i for pure imaginaries, we make the following conventions:

When we are working in a system of algebra requiring the use of complex numbers,

(i) the symbol a will be used as an abbreviation for the number-pair $[a, 0]$,

(ii) the symbol ib will be used as an abbreviation for the number-pair $[0, b]$.

It follows that the symbol $a+ib$ may be used for the number-pair $[a, 0]+[0, b]$, or $[a, b]$.

In particular, we write 1 for the unit $[1, 0]$ of real numbers, and $i\times 1$ for the unit $[0, 1]$ of pure imaginary numbers; but no confusion arises if we abbreviate $i\times 1$ to the symbol i itself.

We do not propose to go into much greater detail, but the reader may easily check that the complex numbers defined in this way by number-pairs have all the properties tentatively proposed for them in the earlier paragraphs of this chapter. We ought, however, to verify explicitly that the square of the number-pair i is the real number -1. For this, we appeal directly to the definition of multiplication:

$$[a, b]\times[u, v] = [au-bv, bu+av].$$

Write $a = u = 0$, $b = v = 1$. Then

$$[0, 1]\times[0, 1] = [-1, 0]$$

or, in terms of the abbreviated symbols,

$$i\times i = -1.$$

Finally, we remark that, although the ordinary language of real numbers is retained without apparent change, the symbols in complex algebra carry an entirely different meaning. For example, the statement

$$2\times 2 = 4$$

remains true; but what we really mean is that

$$\begin{aligned}[2, 0]\times[2, 0] &= [2\times 2-0\times 0, 0\times 2+2\times 0]\\ &= [4, 0].\end{aligned}$$

From now on, however, we shall revert to the normal symbolism without square brackets, writing a complex number in the form $a+ib$ as before. It must always be remembered that number-pairs are intended.

7. The Argand diagram. Abstractly viewed, a complex number $a+ib$ is a 'number pair' $[a, b]$ in which the first component of the pair corresponds to the real part and the second component to the imaginary. A number-pair is, however, also capable of a familiar *geometrical* interpretation, when the two numbers a, b in assigned order are taken to be the Cartesian coordinates of a point in a plane. We therefore seek next to link these two conceptions of number-pair so that we can incorporate the ideas of analytical geometry into the development of complex algebra.

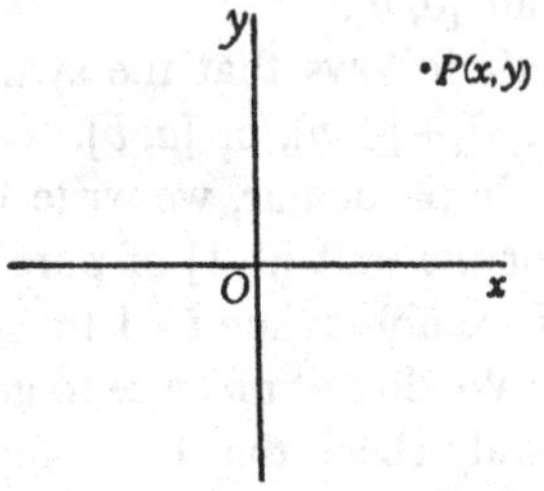

With a change of notation, denote a complex number by the letter z, where

$$z \equiv x+iy.$$

We do this to strengthen the implication of the real and imaginary parts x and y as the Cartesian coordinates of a point $P(x, y)$ (fig. 2).

Fig. 2

There is then an exact correspondence between the complex numbers $z \equiv x+iy$ and the points $P(x, y)$ of the plane:

(i) If z is given, then its real and imaginary parts are determined and so P is known;

(ii) If P is given, then its coordinates are determined, and so z is known.

We say that P *represents* the complex number z. The diagram in which the complex numbers are represented is called an *Argand diagram*, and the plane in which it is drawn is called the *complex plane*, or, when precision is required, the *z-plane*.

Examples 6

Mark on an Argand diagram the points which represent the following complex numbers:

1. $3+4i$. **2.** $5-i$. **3.** $6i$.

4. -3. **5.** $1+i$. **6.** $-2-3i$.

7. $\cos\theta+i\sin\theta$ for $\theta = 0°, 30°, 60°, \ldots, 330°, 360°$.

8. $2\cos^2\frac{1}{2}\theta+i\sin\theta$ for $\theta = 0, 45°, 90°, \ldots, 315°, 360°$.

8. Modulus and argument. Given the point $P(x, y)$ representing the complex number

$$z \equiv x+iy,$$

we may describe its position alternatively by means of polar coordinates r, θ (fig. 3), where r is the distance OP, essentially positive, and θ is the angle $\angle xOP$ which OP makes with the x-axis, measured in the counter-clockwise sense from Ox. Thus r is defined uniquely, but θ is ambiguous by multiples of 2π.

We know from elementary trigonometry that

$$x = r\cos\theta, \quad y = r\sin\theta,$$

whatever the quadrant in which P lies.

Squaring and adding these relations, we have

$$r^2 = x^2+y^2,$$

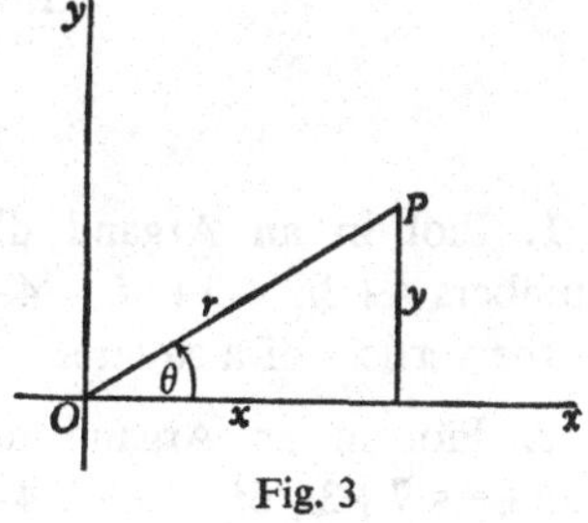

Fig. 3

so that, since r is positive,

$$r = +\sqrt{(x^2+y^2)}.$$

Dividing the relations, we have

$$\tan\theta = y/x,$$

or

$$\theta = \tan^{-1}(y/x).$$

The choice of quadrant for θ depends on the signs of both y and x; if x, y are $+, +$, the quadrant is the first; if x, y are $-, +$, the second; if x, y are $-, -$, the third; if x, y are $+, -$, the fourth.

The two numbers

$$r = \sqrt{(x^2+y^2)},$$
$$\theta = \tan^{-1}(y/x),$$

with appropriate choice of the quadrant for θ, are called the *modulus* and *argument* respectively of the complex number z. If r, θ are given, then

$$z = r(\cos\theta + i\sin\theta).$$

We have already (p. 26) explained the notation $|z| = r$, and proved the formula $z\bar{z} = |z|^2$.

***Illustration* 1.** *To find the modulus and argument of the complex number* $-1+i\sqrt{3}$.

If the modulus is r, then

$$r^2 = (-1)^2+(\sqrt{3})^2 = 1+3 = 4,$$

so that $r = 2$. The number is therefore

$$2\left\{-\frac{1}{2}+i\frac{\sqrt{3}}{2}\right\},$$

so that, if θ is the argument,

$$\cos\theta = -\tfrac{1}{2}, \quad \sin\theta = \frac{\sqrt{3}}{2}.$$

Hence $\theta = \tfrac{2}{3}\pi$, and so the number is

$$2\{\cos\tfrac{2}{3}\pi + i\sin\tfrac{2}{3}\pi\}.$$

Examples 7

1. Plot in an Argand diagram the points which represent the numbers $4+3i$, $-3+4i$, $-4-3i$, $3-4i$, and verify that the points are at the vertices of a square.

2. Plot in an Argand diagram the points which represent the numbers $7+3i$, $6i$, $-3-i$, $4-4i$, and verify that the points are at the vertices of a square.

3. Find the modulus and argument in degrees and minutes of each of the complex numbers

$$\sqrt{3}-i, \quad 3+4i, \quad -5+12i, \quad 3, \quad 6-8i, \quad -2i.$$

4. Prove that, if the point z in the complex plane lies on the circle whose centre is the (complex) point a and whose radius is the real number k, then

$$|z-a| = k.$$

5. Find the locus in an Argand diagram of the point representing the complex number z subject to the relation

$$|z+2| = |z-5i|.$$

9. The representation in an Argand diagram of the sum of two numbers. Suppose that

$$z_1 \equiv x_1 + iy_1, \quad z_2 \equiv x_2 + iy_2$$

are two complex numbers represented in an Argand diagram by the points

$$P_1(x_1, y_1), \quad P_2(x_2, y_2)$$

respectively (fig. 4).

Their sum is the number

$$\begin{aligned} z &= z_1 + z_2 \\ &= (x_1+x_2) + i(y_1+y_2) \end{aligned}$$

represented by the point $P(x, y)$, where

$$x = x_1 + x_2, \quad y = y_1 + y_2.$$

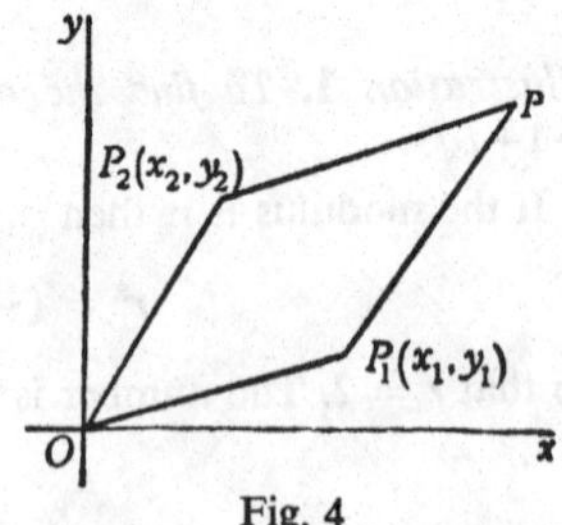

Fig. 4

By elementary analytical geometry, the lines P_1P_2 and OP have the same middle point $\left(\frac{x_1+x_2}{2}, \frac{y_1+y_2}{2}\right)$, so that OP_1PP_2 is a parallelogram. Hence P *(representing* z_1+z_2*) is the fourth vertex of the parallelogram of which* OP_1, OP_2 *are adjacent sides.*

In particular, OP *is the modulus* $|z_1+z_2|$ *of the sum of the two numbers* z_1, z_2.

10. The representation in an Argand diagram of the difference of two numbers. To find the point representing the difference

$$z \equiv z_2 - z_1,$$

we proceed as follows:

The relation is equivalent to:

$$z_2 = z + z_1,$$

so that OP_2 (fig. 5) is the diagonal of the parallelogram of which OP_1, OP (where P represents z) are adjacent sides.

Hence $\overrightarrow{OP}$ *is the line through O parallel to* $\overrightarrow{P_1P_2}$, *the sense along the lines being indicated by the arrows, where* OP, P_1P_2 *are equal in magnitude.*

COROLLARY. *If* $P_1(x_1, y_1)$, $P_2(x_2, y_2)$ *represent the two complex numbers*

$$z_1 \equiv x_1 + iy_1, \; z_2 \equiv x_2 + iy_2,$$

then the line $\overrightarrow{P_1P_2}$ *represents the difference* $z_2 - z_1$, *in the sense that the length* $\overrightarrow{P_1P_2}$ *is the modulus* $|z_2 - z_1|$ *and the angle which* $\overrightarrow{P_1P_2}$ *makes with the x-axis is the argument of* $z_2 - z_1$.

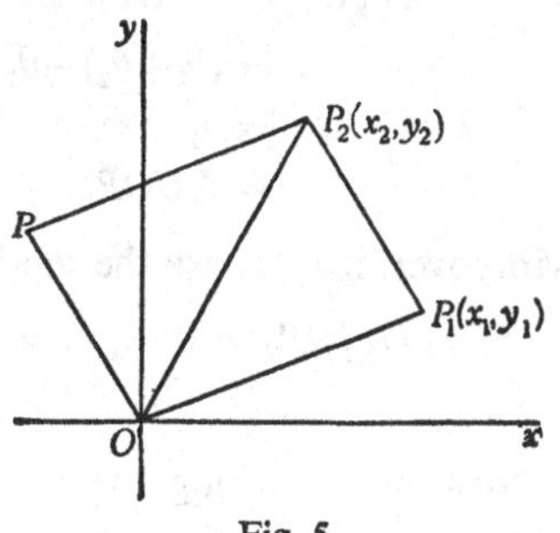

Fig. 5

This corollary is very important in geometrical applications.

[The reader familiar with the theory of vectors should compare the results for the addition and subtraction of two vectors.]

11. The product of two complex numbers. In dealing with the multiplication of complex numbers, it is often more convenient to express them in modulus-argument form. We therefore write

$$z_1 \equiv r_1(\cos\theta_1 + i\sin\theta_1), \quad z_2 \equiv r_2(\cos\theta_2 + i\sin\theta_2).$$

Then
$$\begin{aligned} z_1z_2 &= r_1r_2(\cos\theta_1+i\sin\theta_1)(\cos\theta_2+i\sin\theta_2) \\ &= r_1r_2\{(\cos\theta_1\cos\theta_2-\sin\theta_1\sin\theta_2) \\ &\qquad +i(\sin\theta_1\cos\theta_2+\cos\theta_1\sin\theta_2)\} \\ &= r_1r_2\{\cos(\theta_1+\theta_2)+i\sin(\theta_1+\theta_2)\}, \end{aligned}$$
which is a complex number of modulus r_1r_2 and argument $\theta_1+\theta_2$. Hence

(i) *the modulus of a product is the product of the moduli;*

(ii) *the argument of a product is the sum of the arguments.*

We may obtain similarly the results for division:

(iii) *the modulus of a quotient is the quotient of the moduli;*

(iv) *the argument of a quotient is the difference of the arguments.*

12. The product in an Argand diagram. In the diagram (fig. 6), let P_1, P_2, P represent the two complex numbers z_1, z_2 and their product $z \equiv z_1z_2$ respectively, and let U be the 'unit' point $z = 1$.

Then (§10)
$$\angle xOP = \theta_1+\theta_2,$$
where θ_1, θ_2 are the arguments of z_1, z_2, and so
$$\begin{aligned} \angle P_2OP &= \angle xOP-\angle xOP_2 \\ &= (\theta_1+\theta_2)-\theta_2 \\ &= \theta_1 \\ &= \angle UOP_1. \end{aligned}$$

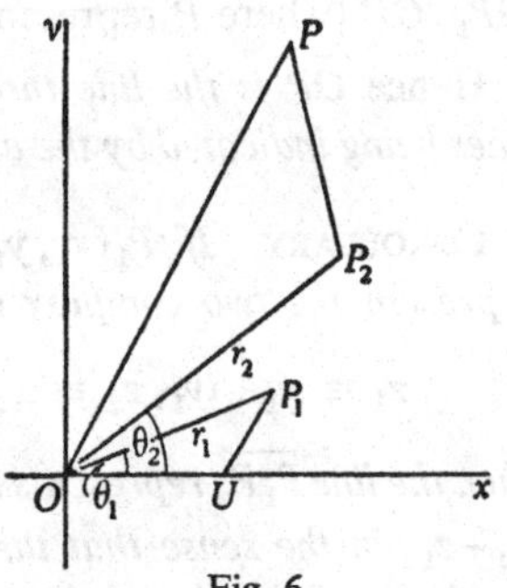

Fig. 6

Moreover, if r_1, r_2 are the moduli of z_1, z_2
$$\begin{aligned} OP/OP_2 &= (r_1r_2)/r_2 = r_1/1 \\ &= OP_1/OU. \end{aligned}$$

Now in measuring the angle $\angle P_2OP$ or $\angle UOP_1$, we proceed, by convention, from the radius vector $\overrightarrow{OP_2}$ or $\overrightarrow{OU}$, through an angle θ_1, in the counter-clockwise sense, to the vector $\overrightarrow{OP}$ or $\overrightarrow{OP_1}$. It may be that θ_1 itself does not lie between $0, \pi$, but nevertheless the fact that the angles $\angle P_2OP$, $\angle UOP_1$ so described are equal ensures that the Euclidean angles $\angle P_2OP$, $\angle UOP_1$ of the triangles ΔP_2OP, ΔUOP_1 are also equal. Moreover, the sides about these equal angles have been proved proportional, and so the triangles are similar.

Hence follows the construction for P:

Let P_1, P_2 represent the two given numbers z_1, z_2; let O be the origin and U the unit point $z = 1$. Describe the triangle POP_2 similar to the *triangle P_1OU, in such a sense that $\angle P_2OP = \angle UOP_1$. Then the point P represents the product z_1z_2.*

COROLLARY. *The effect of multiplying a complex number z by i is to rotate the vector OP representing z through an angle $\frac{1}{2}\pi$ in the counter-clockwise sense.* This follows at once if we put $z_1 = i$ in the preceding work, so that P_1 in the diagram becomes the point (0, 1) or, in polar coordinates, $(1, \frac{1}{2}\pi)$.

Thus i acts in the Argand diagram like an *operator*, turning the radius vector through a right angle.

Examples 8

1. Mark in an Argand diagram the points which represent (*a*) $3+2i$, (*b*) $2+i$, (*c*) their product. Verify from your diagram the theorem of the text.

2. Repeat Ex. 1 for the points

(*a*) 1, (*b*) $3-5i$, (*c*) their product;

(*a*) $2-i$, (*b*) $2+i$, (*c*) their product.

5

THE TRANSFORMATION OF EQUATIONS; SYNTHETIC DIVISION AND THE REMAINDER THEOREM

1. Equations with related roots. When the roots of an equation, say

$$ax^2+bx+c = 0,$$

are known to be p, q, it is useful to be able to form an equation whose roots have some specific relationship to p, q—for example, to form the equation whose roots are twice those of the given equation. This is the problem now to be considered, with attention restricted for the present to *quadratic* equations.

One or two processes are available. We give a general method which is perhaps best exhibited by particular examples.

***Illustration* 1.** *To form the equation whose roots are* -3 *times the roots of the equation*

$$ax^2+bx+c = 0.$$

If the roots of the given equation are p, q, the equation itself expresses a relation in which x can be imagined as replaced by *either* p *or* q. The aim is to form a quadratic equation in a variable, y, say, whose roots are to be $p' \equiv -3p$, $q' \equiv -3q$. This equation will be a relation in which y can be imagined as replaced by *either* p' *or* q'. In other words, the equation is found by writing y (regarded as either p' or q') instead of $-3x$ (regarded as either $-3p$ or $-3q$).

Thus the given equation is transformed from one in x to one in y, where x, y are connected by the relation

$$y = -3x.$$

Substitute $x = -\frac{1}{3}y$ in the given equation; then

$$a(-\tfrac{1}{3}y)^2+b(-\tfrac{1}{3}y)+c = 0,$$

or

$$ay^2-3by+9c = 0.$$

This is the required equation, with the variable denoted by the letter y.

***Illustration* 2.** *To form the equation whose roots are the squares of the roots of the equation*

$$ax^2+bx+c = 0.$$

(The argument is that of the preceding illustration, but expressed with more normal brevity.)

If x is either root of the equation, the aim is to find an equation for y, where

$$y = x^2.$$

The given equation, combined with this relation, gives

$$ay+bx+c = 0,$$

or $$bx = -(ay+c),$$

or, squaring, $$b^2y = (ay+c)^2.$$

Hence the equation is

$$a^2y^2+(2ac-b^2)\,y+c^2 = 0.$$

***Illustration* 3.** *To form the equation whose roots are* $p+1/p$, $q+1/q$, *where* p, q *are the roots of the equation*

$$ax^2+bx+c = 0.$$

If x is either root of the equation, an equation is required for y, where

$$y = x+1/x.$$

Thus $$x^2-yx+1 = 0.$$

Multiply this equation by a and subtract from the first; then

$$(ay+b)x+c-a = 0,$$

so that $$x = \frac{a-c}{ay+b}.$$

Substitute in the given equation, and then multiply throughout by $(ay+b)^2$. Thus

$$a(a-c)^2+b(a-c)(ay+b)+c(ay+b)^2 = 0,$$

or $$acy^2+b(a+c)\,y+\{(a-c)^2+b^2\} = 0,$$

after simplifying and dividing throughout by a. This is the required equation.

Warning Illustration. (Compare p. 59 of the Report *The Teaching of Algebra in Sixth Forms* mentioned on p. 2.)

To form the equation whose roots are the squares of the roots of the equation

$$x^3 = 4x.$$

Write $y = x^2$. Then, since $x^6 = 16x^2$, the required equation is *apparently*

$$y^3 = 16y.$$

BUT the solution $y = -4$ is not the square of any value of x.

In fact, the argument leading to the equation in y is not reversible, since the step $x^6 = 16x^2$ arises not only from the given equation but also from the alternative $x^3 = -4x$. The solution $y = (2i)^2$ therefore intrudes.

More advanced students should train themselves to check their arguments for reversibility. Beginners may well prefer to rest in faith; but they should be especially careful when their manipulation involves steps of this type.

Finally, we give a correct argument, every step of which is reversible:

$$x(x^2-4) = 0,$$

$$\therefore \quad x^2(x^2-4)^2 = 0,$$

so that

$$y(y-4)^2 = 0.$$

Examples 1

Apply the method of Illustrations 1, 2, 3 to solving all the Examples at the end of chapter 3 (p. 22).

Examples 2

(*From examination papers*)

1. The roots of the equation $x^2+px+q = 0$ are α, β. Find the equation whose roots are $\alpha+1, \beta+1$.

2. If α, β are the roots of $x^2+px+q = 0$, and if $\alpha+c, \beta+c$ are the roots of $x^2+Px+Q = 0$, prove that

$$p^2-4q = P^2-4Q.$$

3. The roots of the equation $x^2+x+41 = 0$ are α, β. Without solving this equation, find the equation whose roots are $\alpha-1, \beta-1$.

4. The roots of the equation $ax^2+bx+c = 0$ are α, β. Given that $\alpha = p\beta$, prove that

$$b^2p = ac(1+p)^2.$$

5. Find the equation whose roots are α^2/β and β^2/α, where α, β are the roots of the equation $3x^2-2x+1 = 0$.

6. Given that α, β are the roots of the equation $2x^2-x+2 = 0$, find the equation whose roots are $\alpha+\beta/\alpha, \beta+\alpha/\beta$.

7. Find the equation whose roots are α^2+1, β^2+1, where α, β are the roots of $x^2-6x+2 = 0$.

8. Form the quadratic equation whose roots are the squares of the roots of the equation $x^2+x+2 = 0$.

If α, β are the roots of the given equation, find the value of $\alpha^4+\beta^4$.

9. If α, β are the roots of the equation $x^2+5x+10 = 0$, find the equation whose roots are $2\alpha-3\beta, 2\beta-3\alpha$.

10. If α, β are the roots of $x^2-4x+1 = 0$, prove that $\alpha^3+\beta^3 = 52$, and deduce that $\alpha^6+\beta^6 = 2702$.

11. If α, β are the roots of the equation $x^2-x+5 = 0$, find the equation whose roots are $\alpha^3+1/\beta, \beta^3+1/\alpha$.

12. If α, β are the roots of the equation $x^2-4x+1=0$, find the equation whose roots are α^2+1/α, β^2+1/β.

13. If α, β are the roots of the equation $x^2-5x+7=0$, prove that

$$(\alpha^4+\beta^4)-5(\alpha^3+\beta^3)+7(\alpha^2+\beta^2)=0.$$

Prove also that $\alpha^4+\beta^4=23$.

14. The roots of the equation $ax^2+2hx+b=0$ are α, β. Form the equations whose roots are (i) α^2, β^2, (ii) $\alpha^2+\alpha\beta, \alpha\beta+\beta^2$.

15. Prove that the roots of the equation

$$x^2-2\lambda x+(2\lambda^2-2p\lambda+p^2)=0$$

are not real when λ, p are real unless $\lambda=p$.

16. If α, β are the roots of the equation $ax^2+bx+c=0$, express the the roots of the equation

$$ac(x^2+1)-(b^2-2ac)x=0$$

in terms of α and β.

17. Prove that, if x_1, x_2 are the roots of the equation

$$(x^2+1)(a^2+1)=max(ax+1),$$

then

$$(x_1^2+1)(x_2^2+1)=max_1x_2(x_1+x_2).$$

18. If α, β are the roots of the equation $ax^2+bx+c=0$, express in terms of α, β the roots of the equation

$$acx^2+b(c+a)x+(a+c)^2=0.$$

19. If the roots of the quadratic $ax^2+bx+c=0$ are α, β, obtain in terms of α, β the roots of the equation

$$ax^2+bx+c=\frac{2(b^2-4ac)}{a}.$$

20. If the expression $ax^2+2bx+c$ can be written in the form

$$A(x-x_1)^2+B(x-x_2)^2,$$

where A, B are independent of x, prove that

$$ax_1x_2+b(x_1+x_2)+c=0.$$

2. Quotient and remainder; the method of 'synthetic division'. The work which follows can be applied to a variety of problems, and to equations of any degree. We use for illustration the quartic rather than the quadratic equation, as that makes the 'shape' of the argument clearer.

To find the quotient and remainder when the quartic polynomial

$$ax^4+bx^3+cx^2+dx+e$$

is divided by $x-h$ until the remainder no longer contains x.

By the normal process of division, the quotient is a cubic polynomial with first coefficient a; say

$$ax^3+Bx^2+Cx+D.$$

Suppose that the remainder is R. Then, by definition of quotient and remainder,

$$ax^4+bx^3+cx^2+dx+e \equiv (x-h)(ax^3+Bx^2+Cx+D)+R,$$

an identity, true for all values of x.

The coefficients of the various powers of x on both sides must be equal (an immediate consequence of p. 16, § 3, generalized; see also pp. 54, 57 later), and so

$$b = B-ha, \qquad d = D-hC,$$
$$c = C-hB, \qquad e = R-hD.$$

The coefficients B, C, D and the remainder R may thus be calculated successively by the relations

$$B = b+ha, \qquad D = d+hC,$$
$$C = c+hB, \qquad R = e+hD.$$

The calculation can be effected rapidly by the scheme:

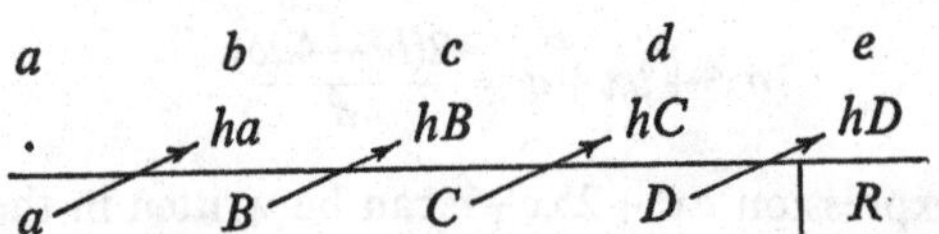

This method of finding quotient and remainder is called *synthetic division.*

***Illustration* 4.** *To find the quotient and remainder on dividing $5x^3-2x^2+1$ by $x-3$.*

The 'line of coefficients' for the given polynomial is

$$5 \quad -2 \quad 0 \quad 1$$

(NOTE. Care must be taken to include zero coefficients.)

Successive stages in the calculation are

(i)	5	−2	0	1
	.	15	?	?
	5	?	?	?

(ii)	5	−2	0	1
	.	15	39	?
	5	13	?	?

(iii)	5	−2	0	1
	.	15	39	117
	5	13	39	?

(iv)	5	−2	0	1
	.	15	39	117
	5	13	39	118

Hence the quotient is $5x^2+13x+39$, and the remainder is 118.

Illustration 5. *To find the quotient and remainder when* $2x^4-3x^3+75x-404$ *is divided by* $x+4$.

[Note that h of the formula is *minus* 4. We now write the solution with normal brevity, save that the arrows remain as guides.]

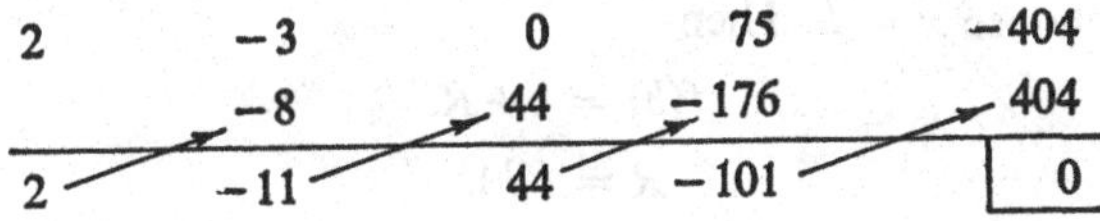

2	−3	0	75	−404
	−8	44	−176	404
2	−11	44	−101	0

The quotient is $2x^3-11x^2+44x-101$ and the remainder 0.

COROLLARY. *Since the remainder is* 0, *the expression* $x+4$ *is a factor of the given polynomial.*

Illustration 6. *To find the quotient and remainder when* $x^3-6x^2+11x+6$ *is divided by* $x+2$.

(Even the arrows are now omitted.)

1	−6	11	6
	−2	16	−54
1	−8	27	−48

The quotient is $x^2-8x+27$ and the remainder -48.

Examples 3

Find the quotient and remainder on dividing

1. x^3-3x^2+4x+7 by $x-2$. **2.** x^3+5x^2+3x+2 by $x-1$.

3. $x^4+4x^3-3x^2+2x+1$ by $x-3$.

4. x^2+5x-4 by $x+2$. **5.** $x^3-6x^2+11x-6$ by $x-2$.

6. $3x^3+2x^2+1$ by $x+1$. **7.** $2x^4+x^2+1$ by $x+2$.

8. $5x^3+x$ by $x+3$. **9.** $4x^3-2x^2-1$ by $x-2$.

10. $7x^4-17$ by $x+2$.

3. The Remainder Theorem. The method of synthetic division acquires increased importance when used in conjunction with the Remainder Theorem. To prove that, *if a polynomial $f(x)$ is divided by $x-h$ until the remainder does not contain x, then the remainder R is equal to the value of the polynomial when x is replaced by h; that is,*

$$R = f(h).$$

Suppose that the quotient is denoted by $g(x)$. Then, by definition of quotient and remainder,

$$f(x) = (x-h)\,g(x)+R.$$

But R does not depend on x, so that its value may be found by giving x any value whatsoever in this identity. In particular (to take the obvious value) we choose $x = h$. Then

$$f(h) = 0+R,$$

so that
$$R = f(h).$$

COROLLARY. *If h is a value of x which makes $f(x)$ zero, then $x-h$ is a factor of $f(x)$.*

For if $f(h) = 0$, then $R = 0$; so that

$$f(x) \equiv (x-h)\,g(x),$$

with $x-h$ as a factor.

The Remainder Theorem. *Alternative proof.* The proof which follows is perfectly general, but is given, for simplicity of statement, in terms of a cubic polynomial.

Suppose that $f(x)$ is the cubic polynomial

$$f(x) \equiv ax^3+bx^2+cx+d.$$

Then
$$f(h) \equiv ah^3+bh^2+ch+d.$$

Hence

$$\begin{aligned} f(x)-f(h) &= a(x^3-h^3)+b(x^2-h^2)+c(x-h) \\ &= a(x-h)(x^2+hx+h^2)+b(x-h)(x+h)+c(x-h) \\ &= (x-h)\{a(x^2+hx+h^2)+b(x+h)+c\} \\ &= (x-h)\{ax^2+(ah+b)x+(ah^2+bh+c)\}, \end{aligned}$$

so that $f(x) = (x-h)\{ax^2+(ah+b)x+(ah^2+bh+c)\}+f(h)$.

Thus, on division of $f(x)$ by $x-h$, the quotient is the quadratic polynomial

$$ax^2+(ah+b)x+(ah^2+bh+c)$$

and *the remainder is* $\quad f(h)$.

This particular proof emphasizes the fact that the operation of the remainder theorem requires $f(x)$ to be a *polynomial.*

NOTE. *If n is a positive integer, then the expression*

$$x^n-a^n$$

contains $x-a$ as a factor.

In fact, the quotient is

$$x^{n-1}+x^{n-2}a+x^{n-3}a^2+\ldots+xa^{n-2}+a^{n-1},$$

since $(x-a)(x^{n-1}+x^{n-2}a+\ldots+a^{n-1})$

$$\begin{aligned} &= x^n+x^{n-1}a+\ldots+xa^{n-1}-x^{n-1}a-x^{n-2}a^2-\ldots-a^n \\ &= x^n-a^n. \end{aligned}$$

***Illustration* 7.** *Given that the remainder is $x+2$ when the polynomial x^3+5x^2+ax+b is divided by x^2-3x+2, to find the values of a, b.*

If $x+2$ is subtracted from x^3+5x^2+ax+b, the resulting polynomial is *exactly* divisible by x^2-3x+2; that is, by $x-1$ and by $x-2$. Thus

$$f(x) \equiv x^3+5x^2+(a-1)x+(b-2)$$

is divisible by $x-1$, $x-2$, so that

$$f(1) = 0, \quad f(2) = 0.$$

Hence $\quad 1+5+(a-1)+(b-2) = 0,$

or $\quad a+b = -3$

and $\quad 8+20+2(a-1)+(b-2) = 0,$

or $\quad 2a+b = -24.$

Thus $\quad a = -21, \quad b = 18.$

4. The value of a polynomial. The Remainder Theorem is often used, as its name implies, as a method for finding the remainder R without going through the tedious process of actual division. An interesting, combination of the results of this chapter, however, provides also a technique for finding rapidly the value assumed by a given polynomial for a given value of the variable x.

Suppose, for example, that the value of the polynomial

$$f(x) \equiv 3x^4+20x^3-6x^2-47$$

is required when $x = -7$. We *could* use the process:

$$\begin{array}{rlrl} 7^4 = & 2401, & 3 \cdot 7^4 = & 7203 \\ 7^3 = & 343, & -20 \cdot 7^3 = & -6860 \\ 7^2 = & 49, & -6 \cdot 7^2 = & -\ \ 294 \\ & & -47 = & -\ \ \ 47 \\ \hline & & & +7203 \\ & & & -7201 \\ \hline & & & 2, \end{array}$$

so that $$f(-7) = 2.$$

But, as an alternative, note that $f(-7)$ is the remainder on dividing $f(x)$ by $x+7$; and this, by synthetic division, is found from the scheme

$$\begin{array}{rrrrr} 3 & 20 & -6 & 0 & -47 \\ & -21 & +7 & -7 & +49 \\ \hline 3 & -1 & 1 & -7 & \boxed{2} \end{array}$$

so that $$f(-7) = 2.$$

***Illustration* 8.** *To find the value of*

$$f(x) \equiv 4x^3-50x^2-27x+7,$$

when $x = 13$.

The value is found from the scheme

$$\begin{array}{rrrr} 4 & -50 & -27 & 7 \\ & 52 & 26 & -13 \\ \hline 4 & 2 & -1 & \boxed{-6} \end{array}$$

so that $$f(13) = -6.$$

Examples 4

1. If $f(x) \equiv x^3+3x^2+2x+1$, find $f(2)$.
2. If $f(x) \equiv x^4+3$, find $f(3)$.
3. If $f(x) \equiv x^4-3x+5$, find $f(-2)$.

4. If $f(x) \equiv 2x^3+3x^2-7$, find $f(-1)$.

5. If $f(x) \equiv 3x^4+2x+9$, find $f(4)$.

6. If $f(x) \equiv 5x^4-2x^3-7$, find $f(-2)$.

7. If $f(x) \equiv x^5+3x+7$, find $f(3)$.

8. If $f(x) \equiv 2x^3-7x-9$, find $f(-2)$.

5. The reduction by a constant of the roots of an equation. Continued application of the idea explained in the preceding paragraph provides a solution for the important problem: *To find the equation whose roots are h less than those of a given equation.*

A particular example will illustrate the argument.

***Illustration* 9.** *To find the equation whose roots are* 3 *less than those of the equation*

$$f(x) \equiv x^3+7x+5 = 0.$$

If y is a typical root of the new equation, then $y = x-3$. Suppose that the new equation is

$$ay^3+by^2+cy+d = 0.$$

Then the equation

$$a(x-3)^3+b(x-3)^2+c(x-3)+d = 0$$

has the same roots as the given equation, so that the polynomial

$$a(x-3)^3+b(x-3)^2+c(x-3)+d$$

is a constant multiple of the polynomial

$$x^3+7x+5.$$

The two polynomials will thus be identically equal if the coefficients of x^3 are taken to be the same; that is, if $a = 1$. Hence $f(x)$ can be written in the form

$$f(x) \equiv (x-3)^3+b(x-3)^2+c(x-3)+d$$

for values of b, c, d to be determined.

Now the right-hand side of this identity can be written

$$(x-3)\{(x-3)^2+b(x-3)+c\}+d,$$

so that d *is the remainder on dividing* $f(x)$ *by* $x-3$, *the quotient being*

$$g(x) \equiv (x-3)^2+b(x-3)+c.$$

Further, $g(x)$ may be written in the form

$$(x-3)\{(x-3)+b\}+c,$$

so that c *is the remainder on dividing* $g(x)$ *by* $x-3$, *the quotient being*

$$h(x) \equiv (x-3)+b.$$

Finally, b *is the remainder on dividing* $h(x)$ *by* $x-3$.

The calculations are now easily effected by synthetic division:

(i) For d:

$$\begin{array}{cccc} 1 & 0 & 7 & 5 \\ & 3 & 9 & 48 \\ \hline 1 & 3 & 16 & \boxed{53} \end{array}$$

Thus
$$d = 53,$$
$$g(x) \equiv x^2+3x+16.$$

(ii) For c:

$$\begin{array}{ccc} 1 & 3 & 16 \\ & 3 & 18 \\ \hline 1 & 6 & \boxed{34} \end{array}$$

Thus
$$c = 34,$$
$$h(x) \equiv x+6.$$

(iii) For b:

$$\begin{array}{cc} 1 & 6 \\ & 3 \\ \hline 1 & \boxed{9} \end{array}$$

Thus
$$b = 9.$$

The whole calculation may be set out more briefly in the form

$$\begin{array}{cccc} 1 & 0 & 7 & 5 \\ & 3 & 9 & 48 \\ \hline 1 & 3 & 16 & \boxed{53} \\ & 3 & 18 & \\ \hline 1 & 6 & \boxed{34} & \\ & 3 & & \\ \hline 1 & \boxed{9} & & \end{array}$$

The required polynomial is
$$y^3+9y^2+34y+53.$$

***Illustration* 10.** *To form the equation whose roots are greater by* 2 *than those of the equation*
$$3x^4+5x^3+9x+15 = 0.$$

The calculation is effected by the scheme:

$$\begin{array}{ccccc} 3 & 5 & 0 & 9 & 15 \\ & -6 & 2 & -4 & -10 \\ \hline 3 & -1 & 2 & 5 & \boxed{5} \\ & -6 & 14 & -32 & \\ \hline 3 & -7 & 16 & \boxed{-27} & \\ & -6 & 26 & & \\ \hline 3 & -13 & \boxed{42} & & \\ & -6 & & & \\ \hline 3 & \boxed{-19} & & & \end{array}$$

The required equation is
$$3y^4-19y^3+42y^2-27y+5 = 0.$$

Illustration **11.** *To form the equation whose roots are* 5 *less than those of the equction*

$$7x^4-6x^3+5x^2-4x+3 = 0.$$

The calculation is effected by the scheme

$$\begin{array}{rrrrr}
7 & -6 & 5 & -4 & 3 \\
 & 35 & 145 & 750 & 3730 \\
\hline
7 & 29 & 150 & 746 & \boxed{3733} \\
 & 35 & 320 & 2350 & \\
\hline
7 & 64 & 470 & \boxed{3096} & \\
 & 35 & 495 & & \\
\hline
7 & 99 & \boxed{965} & & \\
 & 35 & & & \\
\hline
7 & \boxed{134} & & &
\end{array}$$

The required equation is

$$7y^4+134y^3+965y^2+3096y+3733 = 0.$$

It will be noticed how easily this method deals with the high numbers.

Other similar problems may be solved by adaptations of this method. The illustration which follows demonstrates a typical example.

Illustration **12.** *To express the polynomial* x^4+1 *in the form*

$$A+Bx+Cx(x-1)+Dx(x-1)(x-2)+Ex(x-1)(x-2)(x-3).$$

Observe that A is the remainder on dividing by x; B is the remainder on dividing the quotient by $x-1$; C is the remainder on dividing the new quotient by $x-2$; and so on.

Thus $A = 1$, the quotient being x^3. The remainder of the work is contained in the scheme:

$$\begin{array}{rrrr}
1 & 0 & 0 & 0 \\
 & 1 & 1 & 1 \\
\hline
1 & 1 & 1 & \boxed{1} \\
 & 2 & 6 & \\
\hline
1 & 3 & \boxed{7} & \\
 & 3 & & \\
\hline
1 & \boxed{6} & &
\end{array}$$

Hence

$$x^4+1 \equiv 1+x+7x(x-1)+6x(x-1)(x-2)+x(x-1)(x-2)(x-3).$$

Examples 5

1. Find the equation whose roots are 2 less than those of

$$x^3+3x^2+5x+4 = 0.$$

2. Find the equation whose roots are 1 greater than those of

$$x^4-3x^2+2x+1 = 0.$$

3. Find the equation whose roots are 3 less than those of

$$x^4-6x^2+x+1 = 0.$$

4. Find the equation whose roots are 2 greater than those of

$$2x^3+3x^2-7 = 0.$$

5. Find the equation whose roots are 3 greater than those of

$$3x^2+5x+7 = 0.$$

6. Find the equation whose roots are 1 less than those of

$$3x^3+5x^2-8 = 0.$$

6. The multiplication by a constant of the roots of an equation.

To find the equation whose roots are k times those of a given equation.

A particular example will illustrate the argument.

***Illustration* 13.** *To find the equation whose roots are* 3 *times those of the equation*

$$x^4-4x^2+5x-7 = 0.$$

Suppose that y is a typical root of the new equation. Then $y = 3x$, so that $x = y/3$. Thus

$$\left(\frac{y}{3}\right)^4-4\left(\frac{y}{3}\right)^2+5\left(\frac{y}{3}\right)-7 = 0,$$

so that

$$y^4-4.3^2y^2+5.3^3y-7.3^4 = 0.$$

Hence the required equation is

$$y^4-36y^2+135y-567 = 0.$$

More generally, if the given equation is

$$ax^4+bx^3+cx^2+dx+e = 0,$$

then, if $y = kx$, so that $x = y/k$,

$$a\left(\frac{y}{k}\right)^4+b\left(\frac{y}{k}\right)^3+c\left(\frac{y}{k}\right)^2+d\left(\frac{y}{k}\right)+e = 0.$$

Hence the equation whose roots are k times the roots of the given equation is

$$ay^4+bky^3+ck^2y^2+dk^3y+ek^4 = 0.$$

Examples 6

1. Find the equation whose roots are 4 times those of

$$x^2+5x+4 = 0.$$

2. Find the equation whose roots are 3 times those of

$$x^3+2x-7 = 0.$$

3. Find the equation whose roots are -2 times those of

$$x^4+3x^3+x-1 = 0.$$

4. Find the equation whose roots are -1 times those of

$$3x^3-2x^2+x-1 = 0.$$

5. Find the equation whose roots are -10 times those of

$$2x^4+3x^2+x-2 = 0.$$

6. Find the equation whose roots are 10 times those of

$$x^3-5x^2+3x+7 = 0.$$

Revision Examples I

1. When x^3+3x^2+ax+b is dividied by x^2-4, the remainder is $x+16$. Find the values of a, b.

2. Find the values of the constants a, b, c, d that make

$$a(x+1)x(x-1)+bx(x-1)+c(x-1)+d \equiv x^3.$$

3. The remainder, when a polynomial $f(x)$ is divided by $(x-a)(x-b)$, is written in the form $A(x-a)+B(x-b)$. Prove that

$$A = \frac{f(b)}{b-a}, \quad B = \frac{f(a)}{a-b}.$$

When $\alpha x^3+\beta x^2+\gamma x+\delta$ is divided by x^2-1, the remainder is kx, where k is a constant; when it is divided by x^2-4, the remainder is k. Prove that

$$\alpha = -\beta = -\tfrac{1}{4}\gamma = \delta = -\tfrac{1}{3}k.$$

4. Prove that, if two polynomials $P(x)$ and $Q(x)$ have a common linear factor $x-p$, then $x-p$ is also a factor of the polynomial

$$P(x)-Q(x).$$

Hence prove that, if the equations

$$ax^3+4x^2-5x-10 = 0, \quad ax^3-9x-2 = 0$$

have a common root, then $a = 2$ or 11.

5. When

$$px^6+qx^5+2x+3$$

is divided by x^2-1, the remainder is $x+10$. Find the values of p, q.

6. The expression ax^2+bx+c assumes the values 6, 12, 20 when $x = 1, 2, 3$ respectively. For what values of x does this expression vanish?

7. Find the values of a, b, c which make

$$x^4 - ax^3 + bx^2 - cx + 6$$

divisible by $x-1$, $x-2$, $x-3$.

8. Find the remainder when $x^3 + 3x^2 + ax + 2$ is divided by $x+4$, and deduce the value of a in order that the expression may be divisible by $x+4$.

9. Find a value of k which will make

$$2x^4 - x^3 - 11x^2 + kx - 3$$

exactly divisible by $x+1$.

Show that, with this value of k, the above expression is exactly divisible by $(x+1)^2$, and hence solve the equation

$$2x^4 - x^3 - 11x^2 + kx - 3 = 0.$$

10. Find the values of a, b, c that will make the expression

$$x^3 + ax^2 + bx + c$$

equal to zero when either 2 or 3 or 5 is substituted for x.

Find what constant must be added to the expression so that the result may be divisible by $x-1$.

11. What are the values of a, b when

$$x^3 + ax^2 + bx + 125$$

is a perfect cube?

12. What are the values of a, b when

$$x^3 - ax^2 - bx + 25$$

contains both $x-1$ and $x-5$ as factors?

13. Find values of a, b for which the expression

$$x^8 + 2x^7 + 3x^2 + ax + b$$

is divisible without remainder by $x^2 + x - 2$.

14. Use the remainder theorem to find which, if any, of the expressions $x-1$, $x+2$, $x-4$ is a factor of

$$x^3 - x^2 - 16x - 20.$$

15. Find the values of b, c which make $x-1$ and $x+2$ factors of $x^3 + bx + c$.

16. When x^6+Px+Q is divided by $(x-2)(x+3)$, the remainder is $2x+1$. Find the values of P, Q.

17. There is a quartic polynomial, with unit coefficient for x^4, which leaves the same remainder on division by x^2-2x+2 as by x^2-4x+5. It is also divisible without remainder by $(x-2)^2$. Find the polynomial.

18. Find the values of a, b in order that

$$x^4+x^3+x^2+ax+b$$

may be exactly divisible by x^2-3x+2.

19. $f(x)$ is a polynomial of the fifth degree, the coefficient of x^5 being 3. The polynomial leaves the same remainder when divided by x^2+1 or x^2+3x+3. It leaves the remainder $4x+5$ when divided by $(x-1)^2(x+1)$. Find $f(x)$.

20. Find a polynomial $f(x)$ of the ninth degree such that $(x-1)^5$ divides $f(x)-1$ and $(x+1)^5$ divides $f(x)+1$.

Prove that the quotients do not vanish for any real value of x.

6

INTRODUCTION TO THE GENERAL POLYNOMIAL

1. Notation. It is customary to use a 'suffix notation' for the coefficients in a polynomial of general degree. When the degree is n, the polynomial is written

$$a_0x^n + a_1x^{n-1} + a_2x^{n-2} + \ldots + a_n.$$

Note that:

(i) There are $n+1$ terms in the expression.

(ii) The general term (typical term) is a_rx^{n-r}.

(iii) In any term, for example the general term just quoted, the sum of the suffix in the coefficient a_r and the exponent in the power x^{n-r} is equal to n; this simple check ensures accuracy in writing the polynomial down.

We shall see shortly (§ 3) that a polynomial of degree n has n zeros. They may be denoted by a suffix notation, such as

$$\alpha_1, \alpha_2, \alpha_3, \ldots, \alpha_r, \ldots, \alpha_n.$$

2. The results to be proved. We state now the theorems which follow by natural extension from those already given for linear (chapter 2) and quadratic (chapter 3) polynomials. The reader may prefer to regard them for the present as sufficiently obvious and to leave a study of the proofs till later.

The results are:

(i) *A polynomial of degree n which vanishes for $x = \alpha_1, \alpha_2, \ldots, \alpha_n$ (with the α's unequal) can necessarily be expressed in the form*

$$a(x-\alpha_1)(x-\alpha_2)\ldots(x-\alpha_n).$$

(ii) *A polynomial of degree n which vanishes for $n+1$ distinct values $x = \alpha_1, x = \alpha_2, \ldots, x = \alpha_n, x = \alpha_{n+1}$ is necessarily zero identically.*

(iii) *A polynomial of degree n can be determined, and determined uniquely, to have assigned values for $n+1$ distinct values of x.*

To prove these results, it is possible (and profitable) to proceed by induction from the results of the earlier chapters. We give, however, an alternative treatment designed especially to remove the restriction from result (i) that the α's must be unequal.

DEFINITION. A polynomial which can be expressed in the form

$$a(x-h)^p(x-\alpha_1)(x-\alpha_2)\ldots \qquad (h \neq \alpha_1, \alpha_2, \ldots)$$

is said to have $x-h$ as a *multiple factor of multiplicity p*. When the polynomial is equated to zero, the equation has h as a *repeated root* or *multiple root of multiplicity p*.

3. The fundamental theorem of algebra. Let $f(x)$ be a given polynomial of degree n in the variable x. We state, without proof (which is difficult), the fundamental theorem of algebra, that *the equation*

$$f(x) = 0$$

has at least one root (not necessarily real).

Once this theorem has been accepted, we can prove the more detailed theorem: *an equation of degree n has precisely n roots, counted with their correct multiplicities.*

The equation $f(x) = 0$ has at least one root, say a. Thus

$$f(a) = 0,$$

so that, by the remainder theorem, $x-a$ is a factor of $f(x)$. The quotient is a polynomial $g(x)$ of degree $n-1$, where

$$f(x) \equiv (x-a)\,g(x).$$

The equation $f(x) = 0$ is therefore satisfied either when $x = a$, or when $g(x) = 0$.

By the fundamental theorem, the equation

$$g(x) = 0$$

has at least one root, say $x = b$, where b may be equal to a if the equation $f(x) = 0$ has multiple roots. The remainder theorem, as before, leads to an identity

$$g(x) \equiv (x-b)\,h(x),$$

where $h(x)$ is a polynomial of degree $n-2$.

The argument may be continued in this way, yielding a succession of polynomials $k(x)$, $l(x)$, ... of degrees $n-3$, $n-4$, ..., till, finally, a linear polynomial $p(x)$ is reached with a single root. Each step produces one root, so that there are n roots in all, not necessarily all distinct.

The work of p. 55 establishes (subject to acceptance of the fundamental theorem of algebra) the theorem of factorization for a polynomial enunciated on p. 54:

Any polynomial $f(x)$ of degree n can be expressed in the form

$$a_0(x-\alpha_1)(x-\alpha_2)\dots(x-\alpha_n),$$

where a_0 is a constant, and where the n numbers $\alpha_1, \alpha_2, \dots, \alpha_n$ (which may be complex) are not necessarily distinct.

One other general theorem deserves mention here: *When the coefficients of the polynomial $f(x)$ are all real, then, if $a+ib$ is a root of the equation $f(x) = 0$, so also is $a-ib$.*

Since $a+ib$ is a root
$$f(a+ib) = 0.$$

Let the real and imaginary parts of the function $f(u+iv)$ be denoted by U, V, so that
$$f(u+iv) \equiv U+iV.$$

Changing the sign of i on either side yields the corresponding identity
$$f(u-iv) \equiv U-iV.$$

In particular, the values $u = a$, $v = b$ make $U = V = 0$, so that
$$f(a-ib) = 0,$$

and hence $a-ib$ is also a root.

NOTE. If the coefficients of $f(x)$ had not been real, it would not have followed that
$$f(a-ib) \equiv U-iV.$$

Consider, for example, the equation
$$f(x) \equiv 5x^2-12(1+i)x+13 = 0.$$

Writing $x \equiv u+iv$ gives
$$f(u+iv) \equiv 5(u+iv)^2-12(1+i)(u+iv)+13$$
$$\equiv (5u^2-5v^2-12u+12v+13)+i(10uv-12u-12v),$$

so that
$$U \equiv 5u^2-5v^2-12u+12v+13,$$
$$V \equiv 10uv-12u-12v.$$

When $u = 2$, $v = 3$, it follows that $U = 0$, $V = 0$, so that one root of the equation is $2+3i$.

On the other hand,
$$f(u-iv) \equiv 5(u-iv)^2-12(1+i)(u-iv)+13$$
$$\equiv (5u^2-5v^2-12u-12v+13)-i(10uv+12u-12v),$$

which is not $U-iV$; and $2-3i$ is not a solution. The equation with roots $2+3i$, $2-3i$ is, in fact,
$$x^2-4x+13 = 0.$$

If the coefficients in $f(x)$ are all real, then complex values of $\alpha_1, \alpha_2, \ldots, \alpha_n$ occur in pairs like $p \pm iq$, and can be grouped in the form

$$(x-p-iq)(x-p+iq),$$

or

$$(x-p)^2+q^2,$$

where p, q are real. Thus $f(x)$ *may be expressed as the product of a number of linear factors with real coefficients and a number of quadratic factors with real coefficients.*

COROLLARY. *If a polynomial of degree n vanishes for more than n values of x, then it is identically zero.*

For if the polynomial vanishes for a *new* value α_{n+1}, then $a_0 = 0$. This is p. 54, § 2, theorem (ii). Compare also p. 70.

4. The condition for repeated roots. Suppose that the equation

$$f(x) = 0$$

has $x = \alpha$ as a repeated root. Then the polynomial can be expressed in the form

$$f(x) \equiv (x-\alpha)^2 g(x),$$

where the polynomial $g(x)$ may or may not contain $x-\alpha$ as a factor. It follows, by differentiation with respect to x, that

$$\begin{aligned} f'(x) &\equiv (x-\alpha)^2 g'(x)+2(x-\alpha)g(x) \\ &\equiv (x-\alpha)\{(x-\alpha)g'(x)+2g(x)\}. \end{aligned}$$

Hence $f'(x)$ also contains $x-\alpha$ as a factor.

Thus, *for the polynomial equation*

$$f(x) = 0$$

to have $x = \alpha$ *as a repeated root, it is necessary that* $x = \alpha$ *is also a root of the equation*

$$f'(x) = 0.$$

We prove further that this condition is sufficient:

It is given that α is a zero of $f(x)$ and of $f'(x)$. Hence there exists a polynomial $g(x)$ such that

$$f(x) \equiv (x-\alpha)\, g(x),$$

and so

$$f'(x) \equiv (x-\alpha)\, g'(x)+g(x).$$

Since α is a zero of $f'(x)$, we have, by the Remainder Theorem,

$$f'(\alpha) = 0,$$

or

$$g(\alpha) = 0.$$

Hence $x-\alpha$ is a factor of $g(x)$, so that there exists a polynomial $h(x)$ such that

$$g(x) \equiv (x-\alpha)\,h(x).$$

Hence

$$f(x) \equiv (x-\alpha)^2\,h(x),$$

so that α is a repeated zero of $f(x)$.

***Illustration* 1.** *To find whether the equation*

$$f(x) \equiv x^4-8x^3+22x^2-24x+8 = 0$$

has repeated roots.

By differentiation,

$$f'(x) \equiv 4x^3-24x^2+44x-24$$
$$\equiv 4(x-1)(x-2)(x-3).$$

If there is any repeated root, it must be 1, 2 or 3. Now

$$f(1) = -1,$$
$$f(2) = 0,$$
$$f(3) = -1.$$

Hence the only repeated root is 2.

In fact,

$$f(x) \equiv (x-2)^2(x^2-4x+2).$$

5. Symmetric functions and skew-symmetric functions. Let α, β, γ be three given numbers and consider the expression

$$u \equiv \alpha^3+\beta^3+\gamma^3-3\alpha\beta\gamma.$$

If β and γ are interchanged, the expression becomes

$$\alpha^3+\gamma^3+\beta^3-3\alpha\gamma\beta,$$

and is thus unaltered, apart from order (which is irrelevant). The expression is said to be *symmetric* in β and γ. It is, in fact, symmetric in each of the pairs (β, γ), (γ, α), (α, β), and is called a *symmetric function of* α, β, γ. Typical symmetric functions of, say, four variables $\alpha, \beta, \gamma, \delta$ are

$$\alpha+\beta+\gamma+\delta,$$
$$\beta^2\gamma^2+\gamma^2\alpha^2+\alpha^2\beta^2+\alpha^2\delta^2+\beta^2\delta^2+\gamma^2\delta^2,$$
$$\alpha^4+\beta^4+\gamma^4+\delta^4+4\alpha\beta\gamma\delta.$$

By contrast, the functions

$$\alpha^2\beta+\beta^2\gamma+\gamma^2\delta+\delta^2\alpha,$$
$$\alpha^4\beta^3\gamma^2\delta,$$

are not symmetrical.

NOTE. If the linear polynomial $a\alpha+b\beta+c\gamma$ is symmetrical in α, β, γ, then $a = b = c$. If the quadratic polynomial

$$a\alpha^2+b\beta^2+c\gamma^2+2f\beta\gamma+2g\gamma\alpha+2h\alpha\beta$$

is symmetrical in α, β, γ then $a = b = c$ and $f = g = h$. Hence *the general symmetrical polynomials of degrees* 1, 2 *in* α, β, γ *are*

$$A(\alpha+\beta+\gamma) \quad \text{and} \quad A(\alpha^2+\beta^2+\gamma^2)+B(\beta\gamma+\gamma\alpha+\alpha\beta).$$

A function such as

$$v \equiv (\beta-\gamma)(\gamma-\alpha)(\alpha-\beta)$$

is said to be *skew-symmetric* in β, γ since interchange of β and γ yields the function

$$(\gamma-\beta)(\beta-\alpha)(\alpha-\gamma),$$

which is the same as v (apart from order) except that it is *opposite in sign.* The function

$$(\beta-\gamma)(\gamma-\alpha)(\alpha-\beta)(\alpha-\delta)(\beta-\delta)(\gamma-\delta),$$

which is changed in sign but not otherwise when any two of $\alpha, \beta, \gamma, \delta$ are interchanged, is said to be a *skew-symmetric function of* $\alpha, \beta, \gamma, \delta$.

COROLLARY. It follows immediately from the definitions that the product or the quotient of *two skew-symmetric functions is symmetric.*

For example

$$\alpha-\beta, \quad \alpha^2-\beta^2$$

are two skew-symmetric functions of α, β. Their quotient

$$(\alpha^2-\beta^2)/(\alpha-\beta)$$

is the symmetric function $\alpha+\beta$.

Examples 1

Test for symmetry and skew-symmetry the functions:

1. $\alpha^3+\beta^3+\gamma^3$. **2.** $\alpha^2\beta+\beta^2\gamma+\gamma^2\alpha$.

3. $\beta^2\gamma+\beta\gamma^2+\gamma^2\alpha+\gamma\alpha^2+\alpha^2\beta+\alpha\beta^2$.

4. $\beta^2\gamma-\beta\gamma^2+\gamma^2\alpha-\gamma\alpha^2+\alpha^2\beta-\alpha\beta^2$.

5. $(\beta-\gamma)(\gamma-\alpha)(\alpha-\beta)(\alpha+\beta+\gamma)$.

6. $(\beta-\gamma)^2(\gamma-\alpha)^2(\alpha-\beta)^2$.

The following theorem is a little hard at this stage, and may be postponed if desired:

THEOREM. *A polynomial which is skew-symmetric in* α *and* β *contains* $\alpha-\beta$ *as a factor.*

Suppose that the polynomial is a function of, say, four numbers $\alpha, \beta, \gamma, \delta$. It is then (p 8) a sum of terms of which

$$A\alpha^p\beta^q\gamma^r\delta^s$$

is typical. Interchange of α, β gives a polynomial with corresponding term

$$A\beta^p\alpha^q\gamma^r\delta^s.$$

By definition of skew-symmetry, this term must have been present in the given polynomial *but with opposite sign*. That is, the polynomial, in virtue of possessing the term

$$A\alpha^p\beta^q\gamma^r\delta^s$$

must also possess the term $-A\alpha^q\beta^p\gamma^r\delta^s$.

Thus the terms of the polynomial can be grouped in pairs like

$$A(\alpha^p\beta^q\gamma^r\delta^s-\alpha^q\beta^p\gamma^r\delta^s).$$

When p and q are equal, this term is zero. If, say, p is the greater of p, q, the contribution is

$$A\alpha^q\beta^q\gamma^r\delta^s(\alpha^{p-q}-\beta^{p-q}),$$

which, by the Note on p. 45, is equal to

$$A\alpha^q\beta^q\gamma^r\delta^s(\alpha-\beta)(\alpha^{p-q-1}+\alpha^{p-q-2}\beta+\ldots+\beta^{p-q-1}).$$

Hence $\alpha-\beta$ is a factor of all pairs, and therefore a factor of the given polynomial.

6. The symmetric functions of the roots; special cases.*

(i) CUBIC POLYNOMIAL. Let

$$f(x) \equiv ax^3+bx^2+cx+d$$

be a cubic polynomial with zeros α, β, γ. Then (p. 54)

$$\begin{aligned} ax^3+bx^2+cx+d &\equiv a(x-\alpha)(x-\beta)(x-\gamma) \\ &\equiv a\{x^3-(\alpha+\beta+\gamma)x^2+(\beta\gamma+\gamma\alpha+\alpha\beta)x-\alpha\beta\gamma\}. \end{aligned}$$

Hence, equating like coefficients, *the roots of the cubic are connected with the coefficients by the relations*

$$\begin{aligned} \alpha+\beta+\gamma &= -b/a, \\ \beta\gamma+\gamma\alpha+\alpha\beta &= c/a, \\ \alpha\beta\gamma &= -d/a. \end{aligned}$$

* For most of the detailed work of this chapter we confine attention to the cases $n = 3, 4$.

The functions $\alpha+\beta+\gamma,\quad \beta\gamma+\gamma\alpha+\alpha\beta,\quad \alpha\beta\gamma$
are called the *elementary symmetric functions* of the roots. It can be proved that any other symmetric polynomial may be expressed in terms of them. Thus, for example,

$$\alpha^2+\beta^2+\gamma^2 = (\alpha+\beta+\gamma)^2-2(\beta\gamma+\gamma\alpha+\alpha\beta)$$
$$= \frac{b^2}{a^2}-\frac{2c}{a} = \frac{b^2-2ac}{a^2};$$

$$\alpha^3+\beta^3+\gamma^3-3\alpha\beta\gamma = (\alpha+\beta+\gamma)(\alpha^2+\beta^2+\gamma^2-\beta\gamma-\gamma\alpha-\alpha\beta)$$

(as may be proved by multiplication of the right-hand side), so that

$$\alpha^3+\beta^3+\gamma^3 = 3\alpha\beta\gamma+(\alpha+\beta+\gamma)\{(\alpha+\beta+\gamma)^2-3(\beta\gamma+\gamma\alpha+\alpha\beta)\}$$
$$= 3\left(-\frac{d}{a}\right)+\left(-\frac{b}{a}\right)^3-3\left(-\frac{b}{a}\right)\left(\frac{c}{a}\right)$$
$$= (3abc-b^3-3a^2d)/a^3.$$

The presence of the denominator a in the above expressions is sometimes thought troublesome. The polynomial may be divided if desired by a and the cubic equation taken in the form

$$x^3+px^2+qx+r = 0.$$

Then
$$\alpha+\beta+\gamma = -p,$$
$$\beta\gamma+\gamma\alpha+\alpha\beta = q,$$
$$\alpha\beta\gamma = -r.$$

Sometimes, too, the equation is written

$$x^3-px^2+qx-r = 0,$$

where now
$$\alpha+\beta+\gamma = p,$$
$$\beta\gamma+\gamma\alpha+\alpha\beta = q,$$
$$\alpha\beta\gamma = r.$$

These minor modifications are matters of convenience, but can be very useful.

REMARK. The symmetric function $\beta\gamma+\gamma\alpha+\alpha\beta$ is sometimes written in the form $\alpha\beta+\beta\gamma+\gamma\alpha$. This is misleading. The first term in order of symmetry is the one which is peculiar in α (in this case, peculiar by not having it), namely $\beta\gamma$; the second term is the one without β, namely $\gamma\alpha$; the third is the one without γ, namely $\alpha\beta$.

The point is often trivial, but there are times when error arises from a failure to appppreciate that $\beta\gamma$ is first in order of symmetry. See, for example, the Illustration on p. 68, where the first function in order of symmetry is $(\beta-\gamma)^2$, and it is expressed in terms of α.

***Illustration* 2.** *To solve the equation*

$$x^3+9x^2-39x-36 = 0,$$

given that it has two roots whose sum is 3.

The sum of all three roots is -9, so that the third root is -12. Hence $x+12$ is a factor of the cubic polynomial.

The solution follows by the scheme:

$$\begin{array}{rrrr} 1 & 9 & -39 & -36 \\ & -12 & 36 & +36, \\ \hline 1 & -3 & -3 & \boxed{0} \end{array}$$

so that the other two solutions, being the roots of the quadratic

$$x^2-3x-3 = 0,$$

are

$$\tfrac{1}{2}(3\pm\sqrt{21}).$$

(ii) QUARTIC POLYNOMIAL. Let

$$f(x) \equiv ax^4+bx^3+cx^2+dx+e$$

be a quartic polynomial with zeros $\alpha, \beta, \gamma, \delta$. Then (p. 54)

$$\begin{aligned} ax^4+bx^3+cx^2+dx+e &\equiv a(x-\alpha)(x-\beta)(x-\gamma)(x-\delta) \\ &\equiv a\{x^4-(\alpha+\beta+\gamma+\delta)x^3+(\beta\gamma+\gamma\alpha+\alpha\beta+\alpha\delta+\beta\delta+\gamma\delta)x^2 \\ &\quad -(\beta\gamma\delta+\gamma\alpha\delta+\alpha\beta\delta+\alpha\beta\gamma)x+\alpha\beta\gamma\delta\}. \end{aligned}$$

Hence, equating like coefficients, *the roots of the quartic are connected with the coefficients by the relations*

$$\begin{aligned} \alpha+\beta+\gamma+\delta &= -b/a, \\ \beta\gamma+\gamma\alpha+\alpha\beta+\alpha\delta+\beta\delta+\gamma\delta &= c/a, \\ \beta\gamma\delta+\gamma\alpha\delta+\alpha\beta\delta+\alpha\beta\gamma &= -d/a, \\ \alpha\beta\gamma\delta &= e/a. \end{aligned}$$

The four functions on the left are called the *elementary symmetric functions* of the roots.

Here, again, the form

$$f(x) \equiv x^4+px^3+qx^2+rx+s$$

is found useful. Then

$$\begin{aligned} \alpha+\beta+\gamma+\delta &= -p, \\ \beta\gamma+\gamma\alpha+\alpha\beta+\alpha\delta+\beta\delta+\gamma\delta &= \quad q, \\ \beta\gamma\delta+\gamma\alpha\delta+\alpha\beta\delta+\alpha\beta\gamma &= -r, \\ \alpha\beta\gamma\delta &= \quad s. \end{aligned}$$

As before (p. 61), the signs before p and r are sometimes changed.

Illustration 3. *To find the value of p, and to solve the equation*

$$x^4-8x^3+19x^2+px+2 = 0,$$

given that there are two roots whose sum is equal to the sum of the other two.

Let the roots be α, β, γ, δ. Then

$$\begin{aligned} \alpha+\beta+\gamma+\delta &= 8, \\ \beta\gamma+\gamma\alpha+\alpha\beta+\alpha\delta+\beta\delta+\gamma\delta &= 19, \\ \beta\gamma\delta+\gamma\alpha\delta+\alpha\beta\delta+\alpha\beta\gamma &= -p, \\ \alpha\beta\gamma\delta &= 2. \end{aligned}$$

It is given that

$$\alpha+\delta = \beta+\gamma.$$

Hence, since $\alpha+\beta+\gamma+\delta = 8$,

$$\alpha+\delta = 4, \quad \beta+\gamma = 4.$$

Also the second equation is

$$\beta\gamma+\alpha\delta+(\beta+\gamma)(\alpha+\delta) = 19,$$

so that

$$\beta\gamma+\alpha\delta = 3.$$

Thus $\beta\gamma$, $\alpha\delta$ are two numbers whose sum is 3 and whose product is 2. They are therefore the roots of the quadratic equation

$$t^2-3t+2 = 0,$$

or

$$(t-1)(t-2) = 0;$$

say

$$\beta\gamma = 1, \quad \alpha\delta = 2.$$

Hence β, γ are the roots of the equation

$$x^2-4x+1 = 0,$$

and α, δ are the roots of the equation

$$x^2-4x+2 = 0.$$

The four roots are thus $2\pm\sqrt{3}$, $2\pm\sqrt{2}$.

Finally,

$$-p = \beta\gamma(\alpha+\delta)+\alpha\delta(\beta+\gamma) = 12,$$

so that

$$p = -12.$$

7. The sums of powers of the roots. If $\alpha_1, \alpha_2, \alpha_3, \ldots, \alpha_n$ are the roots (with repetitions where necessary) of the equation

$$a_0x^n+a_1x^{n-1}+\ldots+a_n = 0,$$

then the sum of the kth powers of the roots is denoted by the notation

$$\begin{aligned} s_k &\equiv \sum_{\lambda=1}^{n} \alpha_\lambda^k \\ &\equiv \alpha_1^k+\alpha_2^k+\ldots+\alpha_n^k. \end{aligned}$$

Thus

$$\begin{aligned} s_1 &\equiv \alpha_1+\alpha_2+\ldots+\alpha_n, \\ s_2 &\equiv \alpha_1^2+\alpha_2^2+\ldots+\alpha_n^2. \end{aligned}$$

Note that

$$s_0 = n.$$

(i) THE CUBIC $\qquad x^3+px^2+qx+r = 0.$

Let α, β, γ be the roots. Then there are three equations like

$$\alpha^3+p\alpha^2+q\alpha+r = 0.$$

Multiply by α^{k-3} (where $k \geqslant 3$):

$$\alpha^k+p\alpha^{k-1}+q\alpha^{k-2}+r\alpha^{k-3} = 0.$$

Similarly,

$$\begin{aligned} \beta^k+p\beta^{k-1}+q\beta^{k-2}+r\beta^{k-3} &= 0, \\ \gamma^k+p\gamma^{k-1}+q\gamma^{k-2}+r\gamma^{k-3} &= 0. \end{aligned}$$

Add:

$$s_k+ps_{k-1}+qs_{k-2}+rs_{k-3} = 0 \quad (k \geqslant 3).$$

Hence *if three successive sums of powers are known, then the next can be calculated by this formula*; and so all powers can be determined when s_0, s_1, s_2 are known. Such an equation is called a *recurrence relation.* Moreover,

$$\begin{aligned} s_0 &= \alpha^0+\beta^0+\gamma^0 = 3, \\ s_1 &= \alpha+\beta+\gamma = -p, \\ s_2 &= \alpha^2+\beta^2+\gamma^2 \\ &= (\alpha+\beta+\gamma)^2-2(\beta\gamma+\gamma\alpha+\alpha\beta) \\ &= p^2-2q. \end{aligned}$$

Hence all sums of powers can be found.

***Illustration* 4.** *To solve the equation*

$$\begin{aligned} x+y+z &= 1, \\ x^2+y^2+z^2 &= 21, \\ x^3+y^3+z^3 &= 55. \end{aligned}$$

Let x, y, z be the roots of the equation

$$t^3+pt^2+qt+r = 0.$$

Then
$$s_1 = -p,$$
so that
$$p = -1.$$
Also
$$s_2 = p^2-2q,$$
so that
$$2q = 1-21 = -20,$$
or
$$q = -10.$$

Finally,
$$s_3+ps_2+qs_1+rs_0 = 0,$$
so that
$$55-21-10+3r = 0,$$
or
$$r = -8.$$

The equation for t is thus

$$t^3-t^2-10t-8 = 0,$$
or
$$(t+1)(t+2)(t-4) = 0.$$

Hence x, y, z are (in any order) the three numbers $-1, -2, 4$.

Illustration 5. *To calculate s_2, s_3, s_4, s_5 for the equation*

$$x^3+ax-b = 0.$$

Let the roots be α, β, γ. Then

$$\alpha+\beta+\gamma = 0,$$
$$\beta\gamma+\gamma\alpha+\alpha\beta = a,$$
$$\alpha\beta\gamma = b.$$

Thus
$$s_0 = \alpha^0+\beta^0+\gamma^0 = 3,$$
$$s_1 = \alpha+\beta+\gamma = 0,$$
$$s_2 = (\alpha+\beta+\gamma)^2-2(\beta\gamma+\gamma\alpha+\alpha\beta)$$
$$= -2a.$$

Now
$$s_r+as_{r-2}-bs_{r-3} = 0,$$
so that
$$s_r = bs_{r-3}-as_{r-2}.$$
Hence
$$s_3 = bs_0-as_1 = 3b,$$
$$s_4 = bs_1-as_2 = 2a^2,$$
$$s_5 = bs_2-as_3 = -2ab-3ab = -5ab.$$

Note the incidental result: *If α, β, γ are three numbers whose sum is zero, then*

$$\frac{\alpha^5+\beta^5+\gamma^5}{5} = \frac{\alpha^3+\beta^3+\gamma^3}{3}\,\frac{\alpha^2+\beta^2+\gamma^2}{2}.$$

(ii) THE QUARTIC $x^4+px^3+qx^2+rx+s = 0.$

Proceeding as for the cubic, we obtain the recurrence relation (the double use of the letter s will not confuse; sums have suffixes):

$$s_k+ps_{k-1}+qs_{k-2}+rs_{k-3}+ss_{k-4} = 0 \quad (k \geqslant 4).$$

***Illustration* 6.** *To calculate s_2, s_3, s_4 for the equation*

$$x^4+px^3+qx^2+rx+s = 0.$$

If $\alpha, \beta, \gamma, \delta$ are the roots, then

$$\alpha+\beta+\gamma+\delta = -p,$$

$$\beta\gamma+\gamma\alpha+\alpha\beta+\alpha\delta+\beta\delta+\gamma\delta = q.$$

Hence, at once,

$$s_2 = (\alpha+\beta+\gamma+\delta)^2-2(\beta\gamma+\gamma\alpha+\alpha\beta+\alpha\delta+\beta\delta+\gamma\delta)$$

$$= p^2-2q.$$

The calculation of s_3 may be undertaken by straightforward computation, but the following device is worth noting:

Since

$$\alpha^4+p\alpha^3+q\alpha^2+r\alpha+s = 0,$$

it follows that

$$\alpha^3+p\alpha^2+q\alpha+r+\frac{s}{\alpha} = 0.$$

Add the four such relations for $\alpha, \beta, \gamma, \delta$:

$$s_3+ps_2+qs_1+rs_0+s\left(\frac{1}{\alpha}+\frac{1}{\beta}+\frac{1}{\gamma}+\frac{1}{\delta}\right) = 0,$$

so that

$$s_3+p(p^2-2q)+q(-p)+r(4)+s(-r/s) = 0,$$

or

$$s_3 = 3pq-p^3-3r.$$

Then

$$\begin{aligned} s_4 &= -ps_3-qs_2-rs_1-ss_0 \\ &= -p(3pq-p^3-3r)-q(p^2-2q)-r(-p)-s(4) \\ &= p^4+4rp+2q^2-4p^2q-4s. \end{aligned}$$

8. General polynomials with given values. The polynomial of degree n with assigned values for $n+1$ distinct values of x can be written down by direct extension of the formulae for linear and quadratic polynomials (p. 16). Suppose that, for $r = 1, 2, \ldots, n+1$, the polynomial has given value A_r at $x = \alpha_r$. It is then

$$\sum_{r=1}^{n+1} A_r \frac{(x-\alpha_1)(x-\alpha_2)\ldots(x-\alpha_{r-1})(x-\alpha_{r+1})\ldots(x-\alpha_{n+1})}{(\alpha_r-\alpha_1)(\alpha_r-\alpha_2)\ldots(\alpha_r-\alpha_{r-1})(\alpha_r-\alpha_{r+1})\ldots(\alpha_r-\alpha_{n+1})}.$$

The multiplier of A_r is formed by the rule:

The numerator is the product of all the linear factors $x-\alpha_1, x-\alpha_2, \ldots, x-\alpha_{n+1}$ *with* $x-\alpha_r$ *omitted;*

The denominator is the same as the numerator with x *replaced by* α_r.

9. Related equations. The method illustrated earlier (p. 38) for finding the equation whose roots bear some assigned relation to the roots of a given quadratic equation can be extended generally. Two Illustrations will show what can be done.

***Illustration* 7.** *To find the sum of the cubes of the roots of the equation*

$$x^3+px^2+qx+r = 0.$$

If α, β, γ are the roots, form first the equation whose roots are $\alpha^3, \beta^3, \gamma^3$. For this, make the substitution

$$x^3 = y.$$

The given equation is then

$$px^2+qx = -x^3-r = -(y+r).$$

Cube both sides:

$$p^3x^6+3p^2qx^5+3pq^2x^4+q^3x^3 = -(y+r)^3,$$

or

$$p^3y^2+3pqy\,(px^2+qx)+q^3y = -(y+r)^3,$$

or

$$p^3y^2-3pqy\,(y+r)+q^3y = -(y+r)^3.$$

Hence

$$y^3+(p^3-3pq+3r)\,y^2+(q^3-3pqr+3r^2)\,y+r^3 = 0.$$

This is the equation with roots $\alpha^3, \beta^3, \gamma^3$.

But the sum of the roots is $-(p^3-3pq+3r)$, so that

$$\alpha^3+\beta^3+\gamma^3 = 3pq-p^3-3r.$$

***Illustration* 8.** *To form the equation whose roots are* $(\beta+\gamma+\delta)^2, (\gamma+\alpha+\delta)^2, (\alpha+\beta+\delta)^2, (\alpha+\beta+\gamma)^2$, *where* $\alpha, \beta, \gamma, \delta$ *are the roots of the equation*

$$x^4-4x^3+2x+7 = 0.$$

Since

$$\alpha+\beta+\gamma+\delta = 4,$$

it follows that

$$\beta+\gamma+\delta = 4-\alpha.$$

The equation with roots $-(\beta+\gamma+\delta)$, $-(\gamma+\alpha+\delta)$, $-(\alpha+\beta+\delta)$, $-(\alpha+\beta+\gamma)$ is thus obtained by the substitution

$$y = x-4.$$

(The minus signs before the roots are inserted merely for convenience.)

It is therefore found by the scheme (p. 48)

1	−4	0	2	7
	4	0	0	8
1	0	0	2	15
	4	16	64	
1	4	16	66	
	4	32		
1	8	48		
	4			
1	12	48	66	15

The equation is thus $y^4+12y^3+48y^2+66y+15 = 0.$

For the equation whose roots are the squares, make the further substitution

$$y^2 = z.$$

Then

$$z^2+12zy+48z+66y+15 = 0,$$

or

$$(z^2+48z+15)^2 = y^2(12z+66)^2$$
$$= z(12z+66)^2$$

or, after reduction,

$$z^4-48z^3+750z^2-2916z+225 = 0.$$

Illustration 9. *The equation of squared differences; reality of roots.* Let α, β, γ be the roots of the cubic equation

$$x^3+px+q = 0.$$

To form the equation whose roots are

$$(\beta-\gamma)^2, \quad (\gamma-\alpha)^2, \quad (\alpha-\beta)^2.$$

The function $(\beta-\gamma)^2$ may be expressed* in terms of α by the relation

$$(\beta-\gamma)^2 = (\beta+\gamma)^2-4\beta\gamma$$
$$= \{(\alpha+\beta+\gamma)-\alpha\}^2-4(\alpha\beta\gamma)/\alpha$$
$$= \alpha^2+4q/\alpha.$$

Make therefore the substitution

$$y = x^2+4q/x,$$

so that

$$x^3-yx+4q = 0.$$

Subtract this equation from the given cubic in x:

$$(y+p)x-3q = 0,$$

or

$$x = \frac{3q}{y+p}.$$

* See the Remark on pp. 61–2.

Substitute in the given equation:

$$\frac{27q^3}{(y+p)^3}+\frac{3pq}{(y+p)}+q = 0.$$

Divide by q (assumed not zero) and multiply by $(y+p)^3$:

$$(y+p)^3+3p(y+p)^2+27q^2 = 0,$$

or

$$y^3+6py^2+9p^2y+4p^3+27q^2 = 0.$$

To examine the reality of the roots of the given equation. It is assumed that p, q are real.

From the equation in y, it follows that

$$(\beta-\gamma)^2(\gamma-\alpha)^2(\alpha-\beta)^2 = -(4p^3+27q^2).$$

Two deductions are immediate:

(i) *If two of the roots are equal, then*

$$4p^3+27q^2 = 0.$$

(ii) *If all the roots are real and distinct, then the right-hand side is positive, so that*

$$4p^3+27q^2 < 0.$$

Consider next the converse statements:

(i) If

$$4p^3+27q^2 = 0,$$

then

$$(\beta-\gamma)^2(\gamma-\alpha)^2(\alpha-\beta)^2 = 0,$$

and so *the roots are not all different.*

(ii) If

$$4p^3+27q^2 < 0,$$

then

$$(\beta-\gamma)^2(\gamma-\alpha)^2(\alpha-\beta)^2 > 0.$$

Suppose, for the moment, that the roots (necessarily all different) are not all real. Since complex roots occur (p. 56) in conjugate pairs, take α real and write $\beta = u+iv$, $\gamma = u-iv$ with u, v real. Then

$$\begin{aligned}(\gamma-\alpha)(\alpha-\beta) &= -(\alpha-\gamma)(\alpha-\beta)\\ &= -(\alpha-u+iv)(\alpha-u-iv)\\ &= -\{(\alpha-u)^2+v^2\},\end{aligned}$$

so that

$$(\gamma-\alpha)^2(\alpha-\beta)^2 = \{(\alpha-u)^2+v^2\}^2.$$

Also

$$\beta-\gamma = 2iv,$$

so that

$$(\beta-\gamma)^2 = -4v^2.$$

Hence

$$\begin{aligned}(\beta-\gamma)^2(\gamma-\alpha)^2(\alpha-\beta)^2 &= -4v^2\{(\alpha-u)^2+v^2\}^2\\ &= \text{negative}.\end{aligned}$$

This contradicts the earlier condition that the expression is positive, and so *the roots of the given equation are all real.*

Consider, finally, the case when *the roots are not all real.* By what we have just proved,

$$(\beta-\gamma)^2(\gamma-\alpha)^2(\alpha-\beta)^2 = \text{negative},$$

so that

$$4p^3+27q^2 > 0.$$

Conversely, if

$$4p^3+27q^2 > 0,$$

then the roots are not all real; for, if they were, then, by (i) the left-hand side would be negative.

To summarize:

The necessary and sufficient conditions for the roots of the equation

$$x^3+px+q = 0$$

to be (i) not distinct, (ii) all real, (iii) not all real are

$$\text{(i) } 4p^3+27q^2 = 0, \quad \text{(ii) } 4p^3+27q^2 < 0, \quad \text{(iii) } 4p^3+27q^2 > 0.$$

The quantity

$$4p^3+27q^2$$

is called the *discriminant* of the cubic equation.

10. Identically equal polynomials. Suppose that two separate calculations for a polynomial give rise to two apparently different forms of answer; say one calculation giving

$$f(x) \equiv a_0+a_1x+a_2x^2+\ldots+a_nx^n$$

and the other giving

$$g(x) \equiv b_0+b_1x+b_2x^2+\ldots+b_mx^m,$$

of orders m, n respectively. We prove that, *if these two polynomials are identically equal (that is, equal for all values of* x*) then they are of the same order, so that* $m = n$, *and also*

$$a_0 = b_0, \quad a_1 = b_1, \quad a_2 = b_2, \quad \ldots, \quad a_n = b_n.$$

Since

$$f(x) \equiv g(x),$$

it follows that

$$f(x)-g(x) \equiv 0,$$

so that

$$(a_0-b_0)+(a_1-b_1)x+(a_2-b_2)x^2+\ldots \equiv 0,$$

a relation true for all values of x. In particular, suppose that, say, n is the greater (if possible) of m, n. Then the left-hand side is a polynomial of degree n which vanishes for more than n values of x, and so (p. 54) it is identically zero. Equating successive coefficients to zero, we have

$$a_0 = b_0, \quad a_1 = b_1, \quad \ldots, \quad a_m = b_m,$$

$$a_{m+1} = 0, \quad a_{m+2} = 0, \quad \ldots, \quad a_n = 0.$$

Hence all coefficients in the given polynomials are equal.

Warning Illustration. The comparison of $\sin x$ *with a polynomial.* The fact that this apparently trivial theorem contains substance of real importance may perhaps be seen by considering a somewhat related example where the identity is not true.

The function

$$f(x) \equiv \sin x$$

is known to vanish for $x = k\pi$, where k is any integer, positive or negative. Form, then, the polynomial

$$g(x) \equiv (x+m\pi)(x+\overline{m-1}\pi)\ldots(x+\pi)x(x-\pi)(x-2\pi)\ldots(x-n\pi),$$

where m, n are any two positive integers. The two functions $f(x)$, $g(x)$ vanish for all of the $m+n+1$ values $-m\pi, \ldots, n\pi$ of x, however large m, n may be. But they are not identical, for, whatever value is chosen for n, there still remain values, such as $n+1$, for which $g(x) \neq 0$, $f(x) = 0$. The function $\sin x$, in fact, is basically different from the polynomials.

Revision Examples II

1. Solve the equation

$$x^4-4x^3-8x^2-16x-48 = 0,$$

given that there are two roots whose sum is zero.

2. Solve the equation

$$x^3-21x^2+126x-216 = 0,$$

given that the roots are in geometric progression.

3. Form the equation whose roots are $\alpha^3, \beta^3, \gamma^3$, where α, β, γ are the roots of the cubic equation

$$x^3-px^2+qx-r = 0.$$

Hence prove that

$$\beta^3\gamma^3+\gamma^3\alpha^3+\alpha^3\beta^3-(\beta\gamma+\gamma\alpha+\alpha\beta)^3 = \alpha\beta\gamma\{\alpha^3+\beta^3+\gamma^3-(\alpha+\beta+\gamma)^3\}.$$

4. Show that the substitution $x = y-h$, with suitably chosen h, turns the quartic equation

$$x^4-8x^3-17x^2+132x+252 = 0$$

into a quartic equation in y containing neither linear nor cubic terms. Hence solve the equation.

5. Solve the equation

$$x^4-11x^3+28x^2+36x-144 = 0,$$

given that the four roots can be divided into two pairs in such a way that the product of the first pair is minus the product of the second pair.

6. Find the equation whose roots are the squares of those of
$$x^3-3x^2-x+2 = 0.$$

7. Solve the equation
$$x^4-2x^3-x^2+6x-6 = 0,$$
which has two roots whose product is -3.

8. It is given that there is a real number α such that both α and $-\alpha$ are roots of the equation
$$x^4+4x^3-x^2+kx-6 = 0.$$
Find k, and solve the equation.

9. The equation $$x^4-x^3+2x^2-2x+7 = 0$$
has roots $\alpha, \beta, \gamma, \delta$. Find the quartic equations whose roots are

(i) $\beta+\gamma+\delta,\quad \gamma+\alpha+\delta,\quad \alpha+\beta+\delta,\quad \alpha+\beta+\gamma,$

(ii) $\alpha^2, \beta^2, \gamma^2, \delta^2$.

10. Solve the equation
$$x^4-2x^3+x-6 = 0,$$
which has two roots whose product is -3.

11. Solve the equation
$$45x^4-54x^3-98x^2+150x-75 = 0,$$
which has two roots whose sum is zero.

12. Solve the equation
$$16x^4-24x^2+16x-3 = 0,$$
given that it has a multiple root.

13. The equation
$$x^4+ax^3+bx^2+cx+d = 0$$
is such that there are two roots whose sum is equal to the sum of the other two. Prove that
$$a^3-4ab+8c = 0.$$
Find all the roots when $a = 2$, $b = -1$, $c = -2$, $d = -3$.

14. Solve the equation
$$x^4+2x^3-14x^2-11x-2 = 0,$$
given that there are two roots whose product is 2.

15. Show that, if k is a root of the equation

$$x^4+3x^3-6x^2-3x+1 = 0,$$

then so also is $(k-1)/(k+1)$. Express the remaining roots in terms of k, and hence solve the equation.

16. The roots of the equation

$$x^3+px^2+qx+r = 0$$

are α, β, γ. Prove that

$$(\alpha^2-\beta\gamma)(\beta^2-\gamma\alpha)(\gamma^2-\alpha\beta) = rp^3-q^3.$$

17. If α, β, γ are the roots of the equation

$$x^3+px+q = 0,$$

simplify the polynomial

$$(\beta\gamma x+\beta+\gamma)(\gamma\alpha x+\gamma+\alpha)(\alpha\beta x+\alpha+\beta),$$

expressing the coefficients in terms of p and q.

18. There are two roots of the equation

$$x^4-8x^3+19x^2+4\lambda x+2 = 0$$

whose sum is equal to the sum of the other two. Find λ, and solve the equation.

19. Find, in terms of p and q, the cubic equation such that, if α is any one of its roots, then $p\alpha+q$ is also a root.

20. The expression $\quad ax^3+bx^2+cx+d$

is equal to 1, 1, 2, 2 when x takes the values 0, 1, 2, 3 respectively. Find its value when $x = 4$.

21. Prove that the expression

$$f(x) \equiv ax^2+bx+c$$

cannot take the same value for three distinct values of x unless $a = b = 0$. Prove also that it is uniquely determined by the values taken for three distinct values of x.

Find $f(x)$ if

$$f(1) = -3, \quad f(2) = 0, \quad f(3) = 7.$$

22. Write down the cubic polynomial which takes the values A, B, C, D when $x = \alpha_1, \alpha_2, \alpha_3, \alpha_4$ respectively.

Deduce the values of

$$\frac{\alpha_1^m}{(\alpha_1-\alpha_2)(\alpha_1-\alpha_3)(\alpha_1-\alpha_4)}+\frac{\alpha_2^m}{(\alpha_2-\alpha_1)(\alpha_2-\alpha_3)(\alpha_2-\alpha_4)}$$

$$+\frac{\alpha_3^m}{(\alpha_3-\alpha_1)(\alpha_3-\alpha_2)(\alpha_3-\alpha_4)}+\frac{\alpha_4^m}{(\alpha_4-\alpha_1)(\alpha_4-\alpha_2)(\alpha_4-\alpha_3)}$$

for $m = 1, 2, 3$.

23. The roots of the equation

$$x^3+px-q = 0$$

are α, β, γ, and $$s_n = \frac{\alpha^n+\beta^n+\gamma^n}{n}.$$

Express s_5, s_7, s_9, s_{11} in terms of s_2 and s_3.

24. If $\alpha+\beta+\gamma = 0$, express

$$\frac{\alpha^7+\beta^7+\gamma^7}{\alpha^5+\beta^5+\gamma^5}$$

as a quadratic polynomial in α, β, γ.

25. Find the sum of the fourth powers of the roots of the equation

$$x^3+x-1 = 0.$$

26. If

$$\alpha+\beta+\gamma = a,$$
$$\alpha^2+\beta^2+\gamma^2 = b,$$
$$\alpha^3+\beta^3+\gamma^3 = c,$$

find $\alpha\beta\gamma$ and $\alpha^4+\beta^4+\gamma^4$ in terms of a, b, c. Verify that, when $a = 0$, they are respectively $\frac{1}{3}c$ and $\frac{1}{2}b^2$.

27. Given that

$$x+y+z = 3,$$
$$x^2+y^2+z^2 = 5,$$
$$x^3+y^3+z^3 = 7,$$

prove that $$x^4+y^4+z^4 = 9.$$

28. Solve the equations

$$\alpha+\beta+\gamma+\alpha\beta\gamma = 0,$$
$$\beta\gamma+\gamma\alpha+\alpha\beta+1 = 0,$$
$$(\alpha^2+1)(\beta^2+1)(\gamma^2+1) = 20.$$

7

SOLUTION OF EQUATIONS

Equations, like people, combine adherence to type with rugged individuality, and each equation (or set of equations) demands attention to both qualities. The duties of a text-book necessarily concentrate on type, but the wise solver will scan each individual carefully to see whether it has properties of its own leading to particular treatment and economy of labour.

Thus the rules which follow are guides; but it may not be true for any particular example that the rule will give the most convenient solution. On the other hand, the rules, correctly followed, do give solutions, and must be known if fluency is to be acquired.

As in so much of algebra, what is basically necessary is an ability to recognize the class to which a given problem belongs, and, equally important, an ability to manipulate particular examples to bring them into conformity with known classes.

It is assumed that the reader can now deal readily with linear and quadratic equations.

1. Palindromic equations.* An equation such as, say,

$$ax^4+bx^3+cx^2+bx+a = 0,$$

in which coefficients equidistant from either end are equal, may be called *palindromic*. Such an equation can be reduced to one of lower order by dividing throughout by (x raised to the power of the middle† term) and then substituting y for $x+1/x$.

Thus, in the example quoted, divide by x^2, giving

$$a(x^2+1/x^2)+b(x+1/x)+c = 0.$$

Now write $$y = x+1/x,$$

so that $$y^2 = x^2+2+1/x^2,$$

or $$y^2-2 = x^2+1/x^2.$$

* Words like *noon*, *minim*, which read the same when reversed, are called palindromes.

† This rule assumes that the polynomial is of even degree. If not, $x+1$ is a factor and can be removed by division, giving a root $x = -1$.

The equation is thus $a(y^2-2)+by+c = 0,$

or $ay^2+by+(c-2a) = 0.$

The solution of this equation gives two values of y; and substitution of them in the relation

$$y = x+1/x, \quad \text{or} \quad x^2-yx+1 = 0,$$

gives, in all, four values of x.

***Illustration* 1.** *To solve the equation*

$$12x^4-56x^3+89x^2-56x+12 = 0.$$

Divide throughout by x^2:

$$12(x^2+1/x^2)-56(x+1/x)+89 = 0.$$

Write $y = x+1/x,$

so that $y^2-2 = x^2+1/x^2$:

$$12(y^2-2)-56y+89 = 0,$$

or $12y^2-56y+65 = 0,$

or $(2y-5)(6y-13) = 0,$

so that $y = \frac{5}{2}$ or $\frac{13}{6}$.

(i) If $y = \frac{5}{2}$, then $x+1/x = \frac{5}{2},$

so that $2x^2-5x+2 = 0,$

or $(2x-1)(x-2) = 0,$

so that $x = \frac{1}{2}$ or 2.

(ii) If $y = \frac{13}{6}$, then $x+1/x = \frac{13}{6},$

so that $6x^2-13x+6 = 0,$

or $(3x-2)(2x-3) = 0,$

so that $x = \frac{2}{3}$ or $\frac{3}{2}$.

Hence the solutions of the equation are

$$2, \ \tfrac{3}{2}, \ \tfrac{2}{3}, \ \tfrac{1}{2}.$$

Alternative method for quartic. After division throughout by a, the equation of a palindromic quartic assumes the form

$$x^4+mx^3+nx^2+mx+1 = 0.$$

To solve it, assume that it factorizes as the product of two palindromic quadratics

$$(x^2+px+1)(x^2+qx+1);$$

then, on multiplying out,

$$x^4+mx^3+nx^2+mx+1 \equiv x^4+(p+q)x^3+(pq+2)x^2+(p+q)x+1.$$

Hence p, q are given by the relations

$$p+q = m,$$
$$pq+2 = n.$$

Thus p, q are the roots of the equation

$$y^2-(p+q)y+pq = 0,$$

or

$$y^2-my+(n-2) = 0.$$

When p, q have been determined, the solution is immediate.

Illustration 2. *To solve by this method the equation* (*p.* 76)

$$12x^4-56x^3+89x^2-56x+12 = 0.$$

Suppose that the left-hand side is

$$12(x^2+px+1)(x^2+qx+1) \equiv 12\{x^4+(p+q)x^3+(pq+2)x^2+(p+q)x+1\}.$$

Then

$$12(p+q) = -56,$$
$$12(pq+2) = 89,$$

so that

$$p+q = -\tfrac{14}{3}, \quad pq = \tfrac{65}{12}.$$

Hence p, q are the roots of the equation

$$y^2+\tfrac{14}{3}y+\tfrac{65}{12} = 0,$$

or

$$12y^2+56y+65 = 0,$$

or

$$(2y+5)(6y+13) = 0;$$

say

$$p = -\tfrac{5}{2}, \quad q = -\tfrac{13}{6}.$$

The equation is thus $(x^2-\tfrac{5}{2}x+1)(x^2-\tfrac{13}{6}x+1) = 0,$

so that *either*

$$2x^2-5x+2 = 0,$$

giving $x = 2, \tfrac{1}{2}$

or

$$6x^2-13x+6 = 0$$

giving $x = \tfrac{3}{2}, \tfrac{2}{3}$.

The solutions are thus $2, \tfrac{3}{2}, \tfrac{2}{3}, \tfrac{1}{2}$.

2. Equations involving square roots. These equations may be illustrated by an example.

Illustration 3. *To solve the equation*

$$\sqrt{(x+3)}+\sqrt{(x-2)} = \sqrt{(x+19)},$$

where *positive* square roots are to be taken.

These equations must, to begin with, be squared to remove the root signs, and this may involve several stages. In this example, first square both sides:

$$(x+3)+(x-2)+2\surd\{(x+3)(x-2)\} = x+19,$$

or, taking the new square root by itself to one side of the equation,

$$2\surd\{(x+3)(x-2)\} = -x+18.$$

Square again: $$4(x+3)(x-2) = (-x+18)^2,$$

or $$4(x^2+x-6) = x^2-36x+324,$$

or $$3x^2+40x-348 = 0,$$

or $$(3x+58)(x-6) = 0.$$

Hence $$x = -\tfrac{58}{3} \quad \text{or} \quad x = 6.$$

Now it is an essential rule that *all solutions obtained after squaring an equation must be checked by substitution.* Compare p. 2.

Here, the checks are as follows:

(i) If $x = -\frac{58}{3} = -19\frac{1}{3}$, then

$$\surd(x+3) = \surd(-16\tfrac{1}{3}) = \surd(-49/3),$$

$$\surd(x-2) = \surd(-21\tfrac{1}{3}) = \surd(-64/3),$$

$$\surd(x+19) = \surd(-\tfrac{1}{3}).$$

It is possible to argue that the presence of square roots of negative numbers has already ruled this solution out; for the words *positive* and *negative* have no meaning for them. But it is interesting to carry the discussion a little further. Write $k \equiv \surd(-\frac{1}{3})$ and keep to the same determination of the square root throughout; then

$$\surd(x+3) = 7k, \quad \surd(x-2) = 8k, \quad \surd(x+19) = k.$$

The left-hand side of the equation is

$$7k+8k$$

and the right-hand side is $$k,$$

so that $x = -19\frac{1}{3}$ is not a solution.

We have, in fact, solved not only the given equation but others introduced in the process of squaring. For the squaring of the two equations

$$\surd(x+3)+\surd(x-2) = \surd(x+19),$$

and $$\surd(x+3)+\surd(x-2) = -\surd(x+19),$$

gives the same result; and the squaring of the twc equations

$$2\surd\{(x+3)(x-2)\} = -x+18,$$

$$-2\surd\{(x+3)(x-2)\} = -x+18,$$

gives the same result. The solution being considered derives from the choice of signs corresponding to the identities

$$-7k+8k = k, \quad 7k-8k = -k.$$

(ii) If $x = 6$, then

$$\sqrt{(x+3)} = 3, \quad \sqrt{(x-2)} = 2, \quad \sqrt{(x+19)} = 5,$$

and the equation is satisfied in virtue of the identity

$$3+2 = 5.$$

Hence the solution of the equation is $x = 6$.

The two types of equation just discussed are very special, and some more general methods must now be given.

3. Equations with integral roots. Consider equations typified by say, the quartic

$$x^4+bx^3+cx^2+dx+e = 0,$$

where the coefficient of x^4 is unity and b, c, d, e are positive or negative integers or zero. If the (unknown) roots are p, q, r, s, the equation (cf. p. 62) is

$$(x-p)(x-q)(x-r)(x-s) = 0,$$

or, on expansion,

$$x^4-(p+q+r+s)x^3+(qr+rp+pq+ps+qs+rs)x^2$$
$$-(qrs+rps+pqs+pqr)x+pqrs = 0.$$

Corresponding coefficients in the two forms of equation are equal; in particular

$$pqrs = e.$$

This relation leads to the useful rule:

Those roots, if any, which are positive or negative integers are (in their numerical values) factors of e.

The procedure is thus to factorize e; and then, using the remainder theorem (by synthetic division), to check whether any particular factor is a root and to find the corresponding quotient.

Illustration 4. *To solve the equation*

$$x^4-5x^3+5x^2+5x-6 = 0.$$

The factors of 6 are 1, 2, 3, 6, so possible solutions are ± 1, ± 2, ± 3, ± 6. Inspection gives $+1$ at once, so synthetic division is applied:

1	−5	5	5	−6
	1	−4	1	6
1	−4	1	6	0

The last term in the quotient is 6, so that, again, ± 1, ± 2, ± 3, ± 6 are possible solutions. Inspection shows that $+1$ is not a solution, but that -1 is.

Divide again:

$$\begin{array}{rrrr} 1 & -4 & 1 & 6 \\ & -1 & 5 & -6 \\ \hline 1 & -5 & 6 & \underline{|0} \end{array}$$

The new quotient is $x^2-5x+6 \equiv (x-2)(x-3)$.

Hence the roots are $+1, -1, 2, 3$.

The method will also isolate any integral solutions even if there are others which are not integers:

Illustration 5. *To solve the equation*

$$x^3-x^2-x-15 = 0.$$

The factors of 15 are 1, 3, 5, 15, so possible roots are ± 1, ± 3, ± 5, ± 15. The values ± 1 are rejected by inspection. For $x = 3$, the scheme is

$$\begin{array}{rrrr} 1 & -1 & -1 & -15 \\ & 3 & 6 & 15 \\ \hline 1 & 2 & 5 & \underline{|0} \end{array}$$

Hence $x = 3$ is a solution. The quotient gives

$$x^2+2x+5 = 0,$$

or

$$x = -1 \pm 2i.$$

The three roots are thus $3, -1+2i, -1-2i$.

4. Some simple 'selection' rules. One or two elementary rules are available to restrict the number of 'possible' roots that need checking. It is assumed that all coefficients are present, that is, that none are zero.

(i) *If the coefficients are all positive, the equation has no positive roots.*

If, for example, the coefficients a, b, c, d, e in the quartic equation

$$ax^4+bx^3+cx^2+dx+e = 0$$

are all positive, then, if x also is positive, the left-hand side is the sum of positive terms, and so cannot be zero.

(ii) *If the coefficients are alternately positive and negative, the equation has no negative roots.*

If, for example, the numbers p, q, r, s, t are all positive, a typical cubic equation is

$$px^3-qx^2+rx-s = 0$$

and a typical quartic equation is

$$px^4-qx^3+rx^2-sx+t = 0.$$

For a negative value, say $-k$, the respective left-hand sides are

$$-pk^3-qk^2-rk-s,$$
$$pk^4+qk^3+rk^2+sk+t,$$

and these, being respectively sums of terms all negative and terms all positive, cannot be zero.

WARNING. The remark that all coefficients must be present is important. For example, the equation

$$x^3-x+6 = 0$$

has -2 as a root.

For ease of discussion we considered in § 3 the case where the coefficient of the highest power is unity. The method can also be applied where it is not, but there is then the *certainty* that (unless all the coefficients are divisible by the first) there are non-integral solutions—either fractions or complex numbers.

Illustration 6. *To solve the equation*

$$3x^3-16x^2+23x-6 = 0.$$

The factors of 6 are ± 1, ± 2, ± 3, ± 6; and, since the coefficients are alternatively positive and negative, the roots are (p. 80) all positive. Hence if there are integral solutions, they must be 1, 2, 3 or 6. The possibility $x = 1$ is discarded by inspection.

To test for $x = 2$, the scheme is

$$\begin{array}{rrrr} 3 & -16 & 23 & -6 \\ & 6 & -20 & 6 \\ \hline 3 & -10 & 3 & \underline{|0} \end{array}$$

Thus $x = 2$ is one solution, the other solutions being the roots of the quadratic

$$3x^2-10x+3 = 0,$$

or

$$(3x-1)(x-3) = 0,$$

or

$$x = 3, \ \tfrac{1}{3}.$$

Hence the solutions are 2, 3, $\frac{1}{3}$.

5. Fractional roots. It is a small step from the last Illustration to a method for locating roots which are rational fractions of the form p/p', in which p, p' are (positive or negative) integers. The equation with, say, four such roots p/p', q/q', r/r', s/s' is

$$\left(x-\frac{p}{p'}\right)\left(x-\frac{q}{q'}\right)\left(x-\frac{r}{r'}\right)\left(x-\frac{s}{s'}\right) = 0,$$

or
$$(p'x-p)(q'x-q)(r'x-r)(s'x-s) = 0,$$
or
$$p'q'r's'x^4-\ldots+pqrs = 0,$$
where only the first and last terms are important. If the equation is given in the form
$$ax^4+bx^3+cx^2+dx+e = 0,$$
where a, b, c, d, e have no factor in common, then, the original form already having integral coefficients since $p, q, \ldots, r', s'$ are integers, we may take
$$a = p'q'r's', \quad e = pqrs.$$

Thus *the numerators of the roots are factors of e and the denominators of the roots are factors of a.*

***Illustration* 7.** *To solve the equation*
$$3x^3+2x^2+17x-6 = 0.$$

Possible integral solutions are $\pm 1, \pm 2, \pm 3, \pm 6$, and it can be checked that none of them satisfy the equation. Possible rational fractional solutions are $\pm\frac{1}{3}, \pm\frac{2}{3}$. To test for $x = \frac{1}{3}$, the scheme is

$$\begin{array}{cccc} 3 & 2 & 17 & -6 \\ & 1 & 1 & 6 \\ \hline 3 & 3 & 18 & \underline{|0} \end{array}$$

(the multiplier at each step being $\frac{1}{3}$), so that $x = \frac{1}{3}$ is a solution. The other solutions are the roots of the equation
$$3x^2+3x+18 = 0,$$
or
$$x^2+x+6 = 0,$$
or
$$x = \frac{-1\pm\sqrt{(-23)}}{2}.$$
Hence the three solutions are
$$\frac{1}{3}, \quad \frac{-1+i\sqrt{23}}{2}, \quad \frac{-1-i\sqrt{23}}{2}.$$

Examples 1

Solve the following equations in which the *positive* values of the square roots are taken:

1. $\sqrt{(2-x)}+\sqrt{(x+3)} = 3.$ 2. $2\sqrt{x}+\sqrt{(2x+1)} = 7.$

3. $\sqrt{(2x+5)}-\sqrt{(3x-5)} = 2.$

4. $x\sqrt{(2+x)}+2\sqrt{(2-x)} = \sqrt{(8+x^3)}.$

5. $\sqrt{(4x-2)}+\sqrt{(x+1)}-2\sqrt{(7-5x)} = 0.$

6. $\sqrt{(x-5)}+\sqrt{(18-x)} = \sqrt{(x+16)}$.

7. $\sqrt{(2x+61)}-\sqrt{(3x-9)} = 2$.
8. $\sqrt{(x+3)}+\sqrt{(x-5)} = 4$.

9. $\sqrt{(x-16)}+\sqrt{(2x-1)} = \sqrt{(5x-25)}$.

10. $\sqrt{(x-1)}+\sqrt{(x+4)} = \sqrt{(3x+10)}$.

11. $\sqrt{(x+1)}+\sqrt{(x+8)} = \sqrt{(6x+1)}$.

12. $\sqrt{(2+x)}-\sqrt{(2-x)} = x$.

13. $\sqrt{(2x)}+\sqrt{(8x+5)}-\sqrt{(6x+13)} = 0$.

14. $\sqrt{(3x^2-7x-30)}-\sqrt{(2x^2-7x-5)} = x-5$.

15. $\sqrt{(169-x^2)}+\sqrt{(244-x^2)} = 15$.

16. $\sqrt{\left(\frac{x}{1-x}\right)}+\sqrt{\left(\frac{1-x}{x}\right)} = \frac{13}{6}$.

Solve the equations:

17. $18x^4+51x^3-334x^2+51x+18 = 0$.

18. $x^4-x^3-4x^2-x+1 = 0$.

19. $x(x+4)+\frac{1}{x}\left(\frac{1}{x}+4\right) = 10$.

20. $2x^6-9x^5+18x^4-23x^3+18x^2-9x+2 = 0$.

21. $x^4+5x^3+8x^2+5x+1 = 0$.

22. $6x^4+5x^3-38x^2+5x+6 = 0$.

23. $x^4+2x^3+3x^2+2x+1 = 0$.

24. Find the real roots of the equation

$$x^8+1+(x+1)^8 = 2(x^2+x+1)^4.$$

Solve the following equations, given that at least one root is an integer:

25. $35x^3-39x^2-59x+63 = 0$.
26. $x^3-28x+48 = 0$.

27. $3x^3-17x^2+21x-4 = 0$.
28. $x^3-3x^2-16x-12 = 0$.

29. $x^3+3x+36 = 0$.
30. $6x^3-13x^2-10x+24 = 0$.

31. $6x^3-11x^2+6x-1 = 0$.
32. $2x^3-x^2-4x+3 = 0$.

33. $12x^3-16x^2+3x+1 = 0$.
34. $x^3-4x^2-x+4 = 0$.

35. $4x^3+21x^2+29x+6 = 0.$ **36.** $3x^3-19x^2+16x+20 = 0.$

37. $x^4-2x^3-9x^2+2x+8 = 0.$

38. $x^4-15x^2+10x+24 = 0.$

39. $x^4-x^3-31x^2+25x+150 = 0.$

40. $4x^4-12x^3+x^2+12x+4 = 0.$

41. $x^4-2x^3-17x^2+18x+72 = 0.$

42. $6x^4+23x^3-43x^2-192x-144 = 0.$

43. Find whether any of the roots of the equation

$$x^5+8x^4+6x^3-42x^2-19x-2 = 0$$

are integers, and solve it completely.

44. Find the real roots of the equation

$$x^5+2x^4-36x^3-149x^2-232x-336 = 0.$$

8

SOLUTION OF SIMULTANEOUS EQUATIONS

The reader is presumably familiar with the method for solving two simultaneous *linear* equations, but it may be helpful to begin by glancing at the essential ideas behind it.

Consider, for example, the two equations

$$2x+3y = 8,$$
$$7x+5y = 17.$$

The first step is to *eliminate* one of the variables, say y, so as to form an equation in x only. This is usually done by some such steps as the following:

Multiply both sides of the first equation by 5 and both sides of the second by 3:

$$10x+15y = 40,$$
$$21x+15y = 51.$$

Subtract the first of these equations from the second:

$$11x = 11.$$

Hence $$x = 1.$$

From the first given equation,

$$2+3y = 8,$$

so that $$y = 2.$$

Hence $$x = 1, \quad y = 2.$$

There is an alternative way of performing the elimination which is often important:

Use the first equation to express y in terms of x:

$$3y = 8-2x,$$

or $$y = \tfrac{1}{3}(8-2x).$$

Substitute this value of y in the second equation:

$$7x+\tfrac{5}{3}(8-2x) = 17.$$

Multiply by 3 and collect terms:

$$11x = 11,$$

so that $$x = 1,$$

and $$y = \tfrac{1}{3}(8-2) = 2.$$

Hence $$x = 1, \quad y = 2.$$

We now consider some methods for solving two simultaneous equations, assuming that at least one of these is not linear.

1. Two equations of which one is linear. When one of two given simultaneous equations is linear, it is usually best to use that linear equation *either* to express x in terms of y, *or* to express y in terms of x. The value obtained in this way can then be substituted in the second equation, which thus involves one variable only.

Illustration 1. *To solve the equations*

$$2x+3y = 1,$$
$$x^2+2xy+5y^2 = 5.$$

There seems little to choose between x and y, so the first equation may be used to express x in terms of y:

$$x = \tfrac{1}{2}(1-3y).$$

Substitute in the second equation:

$$\tfrac{1}{4}(1-3y)^2+(1-3y)y+5y^2 = 5.$$

Multiply by 4 and remove brackets:

$$1-6y+9y^2+4y-12y^2+20y^2 = 20,$$

or $$17y^2-2y-19 = 0,$$

or $$(y+1)(17y-19) = 0,$$

so that $$y = -1 \quad \text{or} \quad y = \tfrac{19}{17}.$$

To find x, *we must substitute these values in the linear equation* (or we might introduce irrelevant solutions). Thus when $y = -1$, $x = \frac{1}{2}(1+3) = 2$, and when $y = \frac{19}{17}$, $x = \frac{1}{2}(1-\frac{57}{17}) = -\frac{20}{17}$.

Hence the solutions are $$x = 2, \quad y = -1,$$

or $$x = -\tfrac{20}{17}, \quad y = \tfrac{19}{17}.$$

NOTE. If we had substituted, say, $y = -1$ in the *quadratic* equation, we should have had

$$x^2-2x+5 = 5,$$

or $$x^2-2x = 0,$$

so that $$x = 2 \quad \text{or} \quad 0.$$

The solution $x = 2$, $y = -1$ is the relevant one of the two correct ones obtained above. The solution $x = 0$, $y = -1$ satisfies the quadratic equation, but *not* the linear.

2. Two equations each of the type $ax^2+bxy+cy^2 = d$.

Given two equations, say

$$ax^2+bxy+cy^2 = d,$$
$$a'x^2+b'xy+c'y^2 = d',$$

it is best to begin by 'cross-multiplying' to form a quadratic equation for the ratio y/x (or x/y). An example will illustrate the steps.

Illustration 2. *To solve the equations*

$$x^2+2xy+3y^2 = 2,$$
$$3x^2-5xy+7y^2 = 15.$$

Multiply both sides of the first equation by 15, both sides of the second by 2, and then subtract:

$$15(x^2+2xy+3y^2)-2(3x^2-5xy+7y^2) = 0,$$

or $$9x^2+40xy+31y^2 = 0,$$

or $$(x+y)(9x+31y) = 0.$$

Hence $$\textit{either} \quad x = -y \quad \textit{or} \quad x = -\tfrac{31}{9}y.$$

(i) If $x = -y$, the first equation gives

$$y^2-2y^2+3y^2 = 2,$$

or $$y^2 = 1,$$

or $$y = \pm 1.$$

Hence* $$x = \mp 1.$$

(ii) If $x = -\frac{31}{9}y$, the first equation gives

$$\tfrac{961}{81}y^2-\tfrac{62}{9}y^2+3y^2 = 2,$$

or $$y^2 = \frac{81}{323},$$

or $$y = \pm\frac{9}{\sqrt{(323)}}.$$

* The symbol $x = \mp 1$ is meant to denote that, of the two values for x, -1 is taken first (to go with $y = +1$) and $+1$ is taken second (to go with $y = -1$).

Hence
$$x = \mp\frac{31}{\sqrt{(323)}}.$$

There are thus four sets of solutions:

$$x = 1, \quad y = -1, \qquad x = -1, \quad y = 1,$$

$$x = \frac{31}{\sqrt{(323)}}, \; y = \frac{-9}{\sqrt{(323)}}, \qquad x = \frac{-31}{\sqrt{(323)}}, \; y = \frac{9}{\sqrt{(323)}}.$$

3. Three linear equations in three variables.

(There are special points of difficulty, to be considered much later, in connection with these equations. For the time being, the examples are chosen to avoid the pit-falls. The kind of trouble involved may, however, be illustrated by the three simultaneous equations:

$$x+y+z = 3,$$
$$2x+3y+4z = 9,$$
$$3x+4y+5z = 12.$$

The last equation is merely the 'sum' of the first two, and so does not give any fresh information; when it is discarded, there is not enough data to solve uniquely the remaining two equations for three unknowns.

Again, the set

$$x+y+z = 3,$$
$$2x+3y+4z = 9,$$
$$3x+4y+5z = 0$$

is even worse; for the equation obtained by adding the first two is

$$3x+4y+5z = 12,$$

and this contradicts the third of the given equations, so that no solutions are available.)

The method for solution, which is a process of continued elimination, may be demonstrated by a particular example.

***Illustration* 3.** *To solve the equations*

$$x-2y+3z = 14,$$
$$3x+5y+z = -4,$$
$$2x+4y+5z = 9.$$

Eliminate x by subtracting three times the first equation from the second:

$$11y-8z = -46.$$

Eliminate x by subtracting twice the first equation from the third:

$$8y-z = -19.$$

Eliminate z by subtracting eight times the fifth equation from the fourth:

$$-53y = 106.$$

Hence $$y = -2.$$

From the fifth equation, $$z = 8y+19 = 3.$$

From the first equation, $$x = 2y-3z+14 = 1.$$

Hence $$x = 1, \quad y = -2, \quad z = 3.$$

For examples, see pp. 91–2, *numbers* 31–38.

Revision Examples III

1. Solve the equations

$$x+2y = 3, \quad \frac{1}{x}+\frac{1}{y} = 2.$$

2. Solve the equations

$$x+y+1 = 0, \quad 4(x^2-y^2)+7 = 0.$$

3. Solve the equations

$$x+3y = 2, \quad x^2+2y^2+3xy = 0.$$

4. Solve the equations

$$x-2y = 5, \quad x^2+xy+y^2 = 3.$$

5. Solve the equations

$$x-y = 3, \quad x^2+y^2 = 17.$$

6. Solve the equations

$$x+2y = 3, \quad 3x^2+4y^2+12x = 7.$$

7. An examiner found an old question paper with the two simultaneous equations

$$5x+4y = 13,$$

$$2x^2+axy+4y^2 = 24,$$

the figure here printed as a being smudged on the copy. There was also a note that one solution was $x = 1$, $y = b$, the figure printed as b being also smudged. Find the values of b and a, and then find the complete solution of the equations.

8. Solve the equations

$$2x+3y+1 = 0, \quad 4x^2-2xy+y^2+4x = 11.$$

9. Solve the equations
$$x+y = 3, \quad x^2+y^2 = 29.$$

10. Solve the equations
$$2x = 3y = 4z, \quad x^2-y^2+2xy-2xz = 128.$$

11. Solve the equations
$$\frac{x-y}{4} = \frac{z-y}{3} = \frac{2z-x}{1}, \quad x+3y+2z = 4.$$

12. Solve the equations
$$5x-2y = x^2-y^2 = 21.$$

13. Solve the equations
$$x+2y = 3, \quad x^2+3xy+3y^2 = 7.$$

14. Solve the equations
$$x^3+y^3 = 91, \quad x+y = 7.$$

15. Solve the equations
$$3x-y = 7, \quad \frac{4}{x}+\frac{3}{y}+1 = 0.$$

16. Solve the equations
$$\frac{2}{x}+\frac{y}{2} = 1 = \frac{3}{x}+\frac{y}{3}.$$

17. Solve the equations
$$x(y-1) = 8, \quad y(x-1) = 9.$$

18. Solve the equations
$$2x-3y = 1, \quad x^2-y^2+3xy+x-2 = 0.$$

19. Solve the equations
$$x+2y = 1, \quad x^2-2xy+4x-3y = 30.$$

20. Solve the equations
$$x-y = 3, \quad x^3-y^3 = 819.$$

21. Solve the equations
$$x+2y = 1, \quad x^2+xy-y^2+3x+y+8 = 0.$$

22. Solve the equations
$$2x+3y = 1, \quad \frac{2}{x}+\frac{3}{y}+2 = 0.$$

23. Solve the equations

$$x+\frac{1}{y} = 1, \quad x^2-\frac{2x}{y}+\frac{1}{y^2} = \frac{1}{4}.$$

24. Solve the equations

$$x+2y = 4, \quad x^3+8y^3 = 28.$$

25. Solve the equations

$$x^2+2y^2 = 3x^2-xy-y^2 = 27.$$

26. Solve the equations

$$x^2-2xy+y^2 = 9, \quad x^2+3xy-10y^2 = 15.$$

27. Solve the equations

$$3x^2-13xy+9y^2 = 15, \quad 7x^2-5y^2+17 = 0.$$

28. Solve the equations

$$x^2+4y^2 = 25, \quad xy = 6.$$

29. Solve the equations

$$x^2-4xy+4y^2 = 9, \quad 3x^2-2xy+y^2 = 3.$$

30. If

$$x^4+y^4 = 97, \quad x+y = 5,$$

find a quadratic equation satisfied by the product xy, and hence find the real solution of the given equations.

31. Solve the simultaneous equations:

$$\begin{aligned} 3x+4y+5z &= 0, \\ 2x-3y-\ z &= -7, \\ x+4y+4z &= 3. \end{aligned}$$

32. Solve the simultaneous equations:

$$\begin{aligned} x+2y+3z &= 14, \\ 3x+\ y+2z &= 11, \\ 2x+3y+\ z &= 11. \end{aligned}$$

33. Solve the simultaneous equations:

$$\begin{aligned} x+\ y+z &= 6, \\ 2x-\ y+z &= 3, \\ x+4y-z &= 6. \end{aligned}$$

34. Solve the simultaneous equations:

$$\begin{aligned} 3x - y - 5z &= 6, \\ -x + 5y - 3z &= 12, \\ -x - 9y + 25z &= -16. \end{aligned}$$

35. Solve the simultaneous equations:

$$\begin{aligned} 3x + 4y + z &= 1, \\ x + 3y + 2z &= 2, \\ x + 2y + 3z &= 3. \end{aligned}$$

36. Solve the simultaneous equations:

$$\begin{aligned} x + 2y &= -3, \\ 2x + z &= 5, \\ x + 2y + z &= 0. \end{aligned}$$

37. Solve the simultaneous equations:

$$\begin{aligned} 5x + 3y + z &= 5, \\ 4x - 2y + 3z &= 4, \\ 6x + 4y - 5z &= -6. \end{aligned}$$

38. Solve the simultaneous equations:

$$\begin{aligned} 3x + 5y + 7z &= 0, \\ 4x + 3y + 2z &= 0, \\ 8x - 11y + 3z &= 33. \end{aligned}$$

9

THE GRAPH OF A POLYNOMIAL

Valuable information about polynomials may be found by examining their graphs. Although graphs do not, strictly speaking, supply formal proofs for theorems, they are useful as guides to draw attention to characteristic properties.

An acquaintance with the elements of the calculus is presumed, particularly with reference to the idea of *gradient*.

In the discussion we shall assume when necessary that the polynomial has been divided by a constant so that the coefficient of the highest power is unity.

1. Behaviour for large values of x. It is found useful to know how a polynomial $f(x)$ behaves for large values, positive or negative, of the variable x. Consider, as a typical example, the quartic polynomial

$$f(x) \equiv x^4+bx^3+cx^2+dx+e.$$

Divide by x^4; then

$$\frac{f(x)}{x^4} = 1+\frac{b}{x}+\frac{c}{x^2}+\frac{d}{x^3}+\frac{e}{x^4}.$$

The constants b, c, d, e being given, it is always possible to choose x so large that

$$\frac{b}{x}+\frac{c}{x^2}+\frac{d}{x^3}+\frac{e}{x^4}$$

is very small indeed—for example, if x is greater than 1000 times the largest numerically of b, c, d, e, the value of the expression is little, if anything, greater than $\frac{1}{1000}$.

Thus

$$\frac{f(x)}{x^4} = 1$$

approximately; so that *the value of $f(x)$ for numerically large values of x resembles closely that of the first term, namely x^4.*

When $f(x)$ is of even degree, the first term is always positive; when $f(x)$ is of odd degree, the first term is positive for positive x and negative for negative x.

It may be remarked at this point that the graph of the elementary function

$$y = x^n$$

is of the type indicated in the two diagrams (fig. 7), the shape depending on whether n is odd or even.

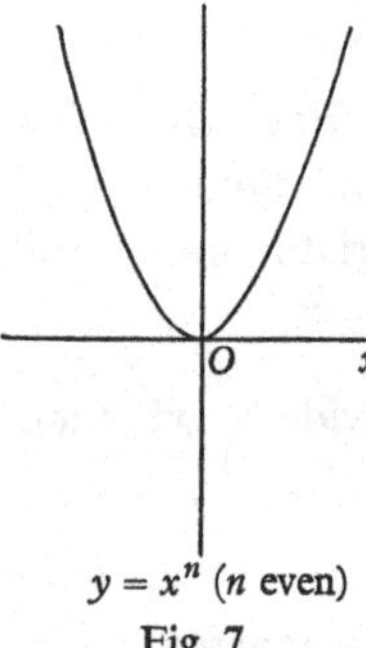

$y = x^n$ (n odd)

$y = x^n$ (n even)

Fig. 7

Examples 1

1. Sketch roughly the curve

$$y = x^n$$

for $n = 1, 2, 3, 4, 5, 6$.

2. Find an approximate value for the ratio:

$$\frac{f(x)}{\text{first term in } f(x)}$$

for the functions

(i) x^3+2x^2+1, (ii) x^4+3x^3-2x-5,

(iii) x^3-x^2+x, (iv) x^4-4x^3+1,

when $x = 1000$ and when $x = 10{,}000$.

2. Features of the graph (roots all different).

Let $f(x)$ be a polynomial such that the roots of the equation

$$f(x) = 0$$

are all different. The diagram (fig. 8) represents a typical curve for the function

$$y = f(x).$$

The behaviour of $f(x)$ as x moves from large negative values to large positive values may be described as follows:

For large negative values of x, $f(x)$ is negative; it increases steadily with x, passing through zero at B, until it reaches a maximum at C; thereafter it decreases steadily, passing through zero at D, to a minimum at E; it then increases again, through zero at F, to a maximum at G, decreases to a minimum at H and then rises steadily as x increases indefinitely.

For reference, denote by $a, b, c, d, \ldots$ the values of x corresponding to the points $A, B, C, D, \ldots$ on the graph.

NOTE. *It is important to observe that, for the moment, the case is excluded when the equation $f(x) = 0$ has equal roots; that is to say, the curve does not touch the x-axis at any point.*

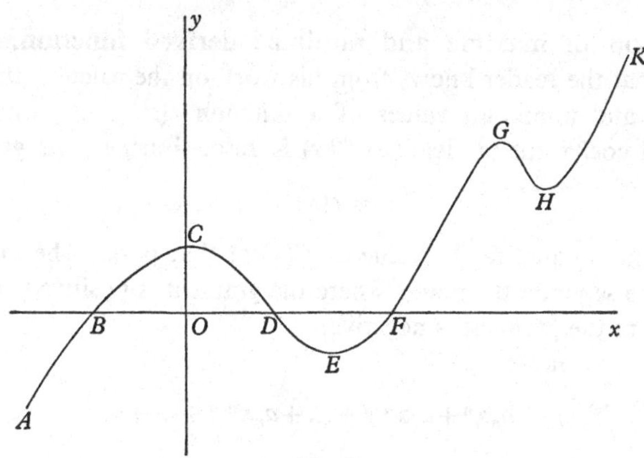

Fig. 8

The reader will readily convince himself (precise mathematical proof may not be easy) that the deductions which follow, although referring to the one particular case, are of general application:

(i) *Maxima and minima occur alternately.* That is to say, if there is a maximum at $x = c$, then as x increases from c, the next turning value (if any) must be a minimum.

(ii) *If $f(x)$ is positive for any value p of x and negative for any value q, then the equation $f(x) = 0$ has an odd number of roots between p, q.*

For example, the graph illustrates a function negative at $x = a$ and positive at $x = k$; there are, in this case, three roots in between.

(iii) *If $f(x)$ is positive (negative) for any value p of x and also positive (negative) for any value q, then the equation $f(x) = 0$ has either no roots or an even number of roots in between.*

For example, the graph illustrates a function negative at $x = a$ and at $x = e$; there are two roots in between. Also $f(x)$ is positive at $x = g$ and at $x = h$, but there are no roots in between.

COROLLARY to (i) and (ii). *An equation of odd degree has at least one real root.*

(iv) *Between consecutive turning values (one maximum and one minimum) the number of roots of the equation is either* 1 *or* 0.

For example, there is one root between $x = c$ and $x = e$; but none between $x = g$ and $x = h$.

(v) *The gradients of $f(x)$ at consecutive zeros have opposite signs.*

Thus, the gradient is positive for $x = b$, negative for $x = d$, positive for $x = f$.

3. Location of maxima and minima; derived function. It is assumed that the reader knows from his work on the calculus that the maximum and minimum values of a function $f(x)$ occur where its differential coefficient (derivative) $f'(x)$ is zero. Further, the gradient of the curve

$$y = f(x)$$

at any point is equal to the value of $f'(x)$ at that point. The maxima and minima separate the points where the gradient is positive from the points where the gradient is negative.

For a polynomial

$$f(x) \equiv a_0x^n + a_1x^{n-1} + \ldots + a_rx^{n-r} + \ldots + a_n,$$

the function

$$f'(x) \equiv na_0x^{n-1} + (n-1)a_1x^{n-2} + \ldots + (n-r)a_rx^{n-r-1} + \ldots + a_{n-1}$$

is also called the *derived function* of $f(x)$.

The *turning points are found by finding the zeros of the derived function and determining which of them are such that the gradient changes sign there.* (When the sign remains unchanged, the curve has an inflexion.)

One formula connected with the derived function is important:

To prove that, if $\alpha_1, \alpha_2, \ldots, \alpha_n$ are the zeros of the polynomial $f(x)$, then

$$f'(x) = \frac{f(x)}{x-\alpha_1} + \frac{f(x)}{x-\alpha_2} + \ldots + \frac{f(x)}{x-\alpha_n}.$$

The polynomial with these zeros is (p. 56)

$$f(x) \equiv a_0(x-\alpha_1)(x-\alpha_2)\ldots(x-\alpha_n).$$

Take logarithms:

$$\log\{f(x)\} = \log a_0 + \log(x-\alpha_1) + \log(x-\alpha_2) + \ldots + \log(x-\alpha_n).$$

Differentiate: $$\frac{f'(x)}{f(x)} = \frac{1}{x-\alpha_1} + \frac{1}{x-\alpha_2} + \ldots + \frac{1}{x-\alpha_n}.$$

***Illustration* 1.** *To find the condition for the equation*

$$x^3 - 12x + k = 0$$

to have three real roots.

If
$$f(x) \equiv x^3 - 12x + k,$$
then
$$f'(x) \equiv 3x^2 - 12 \equiv 3(x^2 - 4).$$
Hence the turning values are at $x = 2$, $x = -2$.

Now
$$f(2) = 8 - 24 + k = k - 16,$$
$$f(-2) = -8 + 24 + k = k + 16.$$

Moreover,
$$f'(x) \equiv 3(x-2)(x+2),$$
so that when x is slightly greater than 2, the sign of $f'(x)$ is $(+)(+)$, or positive; and when x is slightly less than 2, the sign of $f'(x)$ is $(-)(+)$, or negative. Thus as x increases $f'(x)$ passes from negative through zero to positive, so that there is a minimum at $x = 2$.

Similarly, there is a maximum at $x = -2$. The curve
$$y = f(x)$$
is therefore of the type shown (fig. 9), and the number of real roots is the number of times that the curves crosses the x-axis. Thus for three real roots the x-axis must pass between the two points marked A, B. That is, $f(x)$ must be negative at A and positive at B.

Fig. 9

Hence
$$k - 16 = f(2) \quad < 0,$$
$$k + 16 = f(-2) > 0.$$

The condition is therefore that k must lie in the interval
$$-16 < k < 16.$$

4. Features of the graph when the roots are not all different. The diagram (fig. 10) illustrates a case when the polynomial equation
$$f(x) = 0$$
has repeated roots (at $x = c$ and at $x = g$).

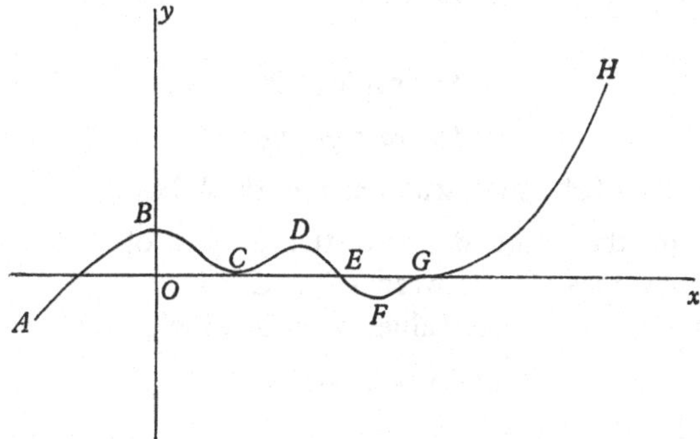

Fig. 10

Suppose that the equation has, say, $x = p$ as a root which is double or more. Then $f(x)$ contains $(x-p)^2$ as a factor, so that it assumes the form

$$f(x) \equiv (x-p)^2\, g(x),$$

where $g(x)$ is the quotient on dividing by $x-p$ twice. If differentiations with respect to x are denoted by dashes, then

$$f'(x) \equiv (x-p)^2 g'(x) + 2(x-p) g(x),$$

so that *the differential coefficient $f'(x)$ also vanishes for $x = p$.*

Hence *the gradient of the curve is zero at a multiple root*; that is, the curve touches the x-axis, which is the tangent at $x = p$.

Suppose, conversely, that the polynomial $f(x)$ is such that its differential coefficient $f'(x)$ vanishes at a value $x = p$ for which $f(x)$ is also zero. Then *the equation $f(x) = 0$ has p as a root which is at least double*:

Suppose that the quotient on dividing $f(x)$ by the factor $x-p$ is $h(x)$, so that

$$f(x) \equiv (x-p) h(x),$$

then
$$f'(x) = (x-p) h'(x) + h(x).$$

Since $f'(p) = 0$, it follows that $h(p) = 0$. Hence, by the Remainder Theorem, $x-p$ is a factor of $h(x)$; say $h(x) \equiv (x-p) k(x)$. Then

$$f(x) \equiv (x-p)^2 k(x),$$

so that $x = p$ is (at least) a double root. (Compare the similar statement on p. 57.)

We do not propose to go into further details at present. When the possibility of repeated roots must be taken into account, it is probably best to draw a rough sketch for the particular case and then to read off the desired properties. Consider, for example, the modifications (if any) in properties (i)–(v) on pp. 95–6.

Examples* 2

Accurate groups

(Take such scales as you think best.)

1. Draw in the same diagram the graphs of $y = x^3 - 2x$ and $y = 1 - x^2$ for values of x between -3 and $+2$.

Find from your graph the values of x for which

$$x^3 + x^2 - 2x - 1 = 0.$$

* Some of the functions are more general than polynomials, but the principles are the same. They are inserted mainly to give practice in accurate graph-work.

2. Draw the graph of $y = x^3+2x$.
Hence find a value of x for which

$$2x^3+4x = 3.$$

3. Draw the graph of $y = \frac{1}{5}(x^3-x+1)$ for values of x from -3 to 3. By drawing a suitable portion of the graph with 5 inches as unit distance on each axis, find, to two places of decimals, the real root of of the equation

$$x^3-x+1 = 0.$$

4. Draw the graph of x^3-3x^2 between the values -1 and 3·5 of x. Hence find approximately the solutions of the equation

$$x^3-3x^2-x+1 = 0.$$

5. Draw in the same diagram the graphs of x^3 and $2x^2+2x-2$ for values of x from -2 to 3.
Hence find to the nearest tenth the solutions of the equation

$$x^3-2x^2-2x+2 = 0.$$

6. By means of the graphs

$$y = x^3, \quad y = 3x+1,$$

find the solutions of the equation

$$x^3-3x-1 = 0.$$

7. Draw the graph of x^3-2x^2 between the values -1 and 2·5 of x. Use the graph to solve the equation

$$2x^3-4x^2-x+1 = 0.$$

8. Draw the graphs of

$$y = x^2-4x, \quad y = x^3-4x^2+1.$$

between the values $x = -1$ and $x = 4$.
Deduce the solutions of the equation

$$(x-1)(x^2-4x)+1 = 0.$$

9. Draw the graph of $y = x^3+x$ from $x = -3$ to $x = +3$.
Use your graph to find the real root of the equation

$$2x^3+3x-4 = 0.$$

10. Draw the graph of

$$y = (x-2)\left(\tfrac{1}{2}+4x-x^2\right)$$

between the values $x = -\frac{1}{2}$ and 4.

Deduce the roots of the equation

$$x-1 = (x-2)\left(\tfrac{1}{2}+4x-x^2\right).$$

11. Draw the graphs of

$$y = \frac{2x-4}{x^2+1}, \quad y = x-3$$

for values of x from -3 to 4, including among the points plotted those for which $x = -\frac{1}{2}, -\frac{1}{4}, +\frac{1}{2}$.

Find the roots of the equation

$$(x^2+1)(x-3)-2x+4 = 0.$$

12. With the same axes and the same scales draw the graphs of the functions

$$\frac{3x^2}{1-x+x^2}, \quad \tfrac{1}{5}x^3,$$

plotting the values of x from -2 to $+3$.

Prove that for all values of x, positive or negative, the first graph lies on the positive side of the x-axis.

Find from your graphs the real root of the equation

$$x(x^2-x+1) = 15.$$

13. Draw the graphs of the functions

$$\frac{x^3}{x^2-16}, \quad x^2-4$$

for values of x between $-2\frac{1}{2}$ and $+2\frac{1}{2}$.

Deduce the values of the two roots, between $-2\frac{1}{2}$ and $+2\frac{1}{2}$, of the equation

$$(x^2-16)(x^2-4)-x^3 = 0.$$

14. Draw the graphs of the functions

$$x^3-3x+2, \quad 3-\tfrac{1}{4}x^2$$

for values of x between -2 and $+2$.

Deduce the roots of the equation

$$4x^3+x^2-12x-4 = 0.$$

15. Draw the graph of the function

$$x^2-3x+2/x$$

for values of x between $\frac{1}{2}$ and 4.

Use your graph to estimate the roots of each of the equations

$$x^3-3x^2+2=0, \quad x^3-3x^2-x+2=0$$

between $x=\frac{1}{2}$ and $x=4$.

16. Draw the graphs of the functions

$$\tfrac{1}{4}x^3, \quad \frac{1}{x+2}$$

for values of x between -1 and $+3$.

Find the positive root of the equation

$$x^4+2x^3=4.$$

17. Draw the graphs of the functions

$$1/x, \quad \tfrac{1}{4}x^2+1$$

for values of x between -2 and 2.

[The first function is undefined ('becomes infinite') for $x=0$.]

Find the positive root of the equation $x^3+4x=4$.

18. Draw the graphs of the functions

$$\tfrac{1}{2}(2-x^3), \quad (x-1)^2$$

for values of x between -1 and $+2$.

Deduce two roots of the equation

$$2-x^3=2(x-1)^2,$$

and show that the accurate value of the positive root is $\sqrt{5}-1$.

19. Draw the graphs of the functions

$$\tfrac{1}{10}x^3, \quad \frac{3x}{x^2+1}$$

for values of x between -4 and $+4$.

Determine from your first graph the cube root of 35.

Prove that the non-zero abscissae of the points of intersection of the two graphs satisfy the equation

$$x^4+x^2-30=0,$$

and hence find from your graphs the square root of 5.

20. Draw the graphs of the functions

$$\frac{x^2-5x+4}{x^2+4}, \quad \tfrac{1}{2}(2-x)$$

for values of x between -4 and $+4$.

Prove that the abscissae of the three points of intersection of the graphs satisfy the equation

$$x^3-6x = 0,$$

and deduce from your graphs the square root of 6.

21. Draw the graph of $y = \frac{1}{8}x^3$ from $x = 0$ to $x = 4$.
Use your graph to estimate the value of (i) $(2{\cdot}8)^3$, (ii) $\sqrt[3]{6}$.

Revision Examples IV

1. By any graphic method show that the equation

$$x^3-3x^2+4x-3 = 0$$

has only one real root.

2. Determine the gradient of the function $(x-1)^2(x-2)$ and the values of x for which the gradient vanishes.

Deduce that the equation

$$x^3-4x^2+5x-4 = 0$$

has only one real root, and that the equation

$$x^3-4x^2+5x-\tfrac{17}{9} = 0$$

has three real roots.

3. Find the gradient of the function

$$(x-1)^2(x-3)^2$$

and the values of x for which this gradient vanishes.

Find for what range of values of n the equation

$$x^4-8x^3+22x^2-24x+n = 0$$

has four real roots.

4. Prove, by graphical methods or otherwise, that a cubic equation (with real coefficients) of the form

$$ax^3+bx^2+cx+d = 0$$

has at least one real root.

Prove that, if λ is real, the equation

$$x^3-3x^2+3x+\lambda = 0$$

is satisfied by one and only one real value of x.

5. Prove, by consideration of a graph or otherwise, that the equation

$$f(x) \equiv a_n x^n + a_{n-1}x^{n-1} + \ldots + a_1 x + a_0 = 0,$$

with real coefficients, cannot have more than $r+1$ real roots when the derived equation $f'(x) = 0$ has exactly r real roots.

Prove that the equation

$$5x^4 - x + 1 = 0$$

has no real root, and deduce that the equation

$$4x^5 - 2x^2 + 4x - 10 = 0$$

has exactly one real root.

Find the number of real roots of the equation

$$x^6 - x^3 + 3x^2 - 15x + 5 = 0.$$

6. Prove that the equation

$$3x^5 - 5ax^3 + b = 0$$

has three real roots when a, b are real and $4a^5 > b^2$.

7. If $f(x)$ is a polynomial in x, show that between consecutive real roots of the derived equation $f'(x) = 0$ there occurs either one root or no root of $f(x) = 0$.

Find the range of values of k for which the equation

$$x^4 - 14x^2 + 24x - k = 0$$

has all its roots real.

8. By means of a graph, determine the values of λ for which the equation

$$(x-1)^2(x-2) + \lambda = 0$$

has three real roots.

9. Draw a rough sketch of the graph of the function

$$y = x^5 - 5x^3 - 5x^2,$$

proving that its two turning values occur when $x = 0$ and when $x = 2$.

Show that the equation

$$x^5 - 5x^3 - 5x^2 + a = 0$$

has three real roots if $0 < a < 28$.

10. Find necessary and sufficient conditions to be satisfied by the coefficients p, q in order that the equation

$$x^3 + px + q = 0$$

may have two positive roots.

10

PARTIAL FRACTIONS

1. The problem. It is a familiar fact that a sum (or difference) of arithmetical fractions, like

$$\tfrac{1}{3}+\tfrac{2}{5}-\tfrac{8}{21}$$

can be combined into the single fraction $\frac{37}{105}$. In the same way, the algebraic expression

$$\frac{x^2}{(x+1)^3}-\frac{2x}{x^2-1}+\frac{x+3}{(x-1)(x^2+1)}$$

can be 'simplified' to the single fraction

$$\frac{-x^5-4x^4+3x^3+7x^2+8x+3}{(x+1)^3(x-1)(x^2+1)}.$$

We must now consider the reverse process. It is often important, as a step in algebraic manipulation, to 'pull apart' a fraction like that just given, so as to exhibit it as a sum (or difference) of simpler fractions. The problem is to settle the precise forms of these simpler fractions, and to give suggestions for calculating them.

There are, in the first instance, many different ways of expressing a decomposed fraction. Thus

$$\begin{aligned}\frac{6x-2x^3}{1-x^4} &\equiv \frac{1}{1-x}-\frac{1-4x-3x^2}{(1+x)(1+x^2)}\\ &\equiv \frac{-1}{1+x}+\frac{1+4x-3x^2}{(1-x)(1+x^2)}\\ &\equiv \frac{1}{1-x}-\frac{1}{1+x}+\frac{4x}{1+x^2},\end{aligned}$$

and so on. The particular decomposition to be exhibited is given the name *partial fractions*. The form, for any one function, is then unique, but there are many ways of performing the calculations, and each student tends to develop a technique of his own. The method which follows has the advantage of incorporating some general theory as it goes, and also of providing, at each stage of the calculation, a check (potent though not quite infallible) on accuracy.

The beginner will presumably learn to calculate the form that he is taught by the method that he is taught. With greater experience he may learn to choose alternative forms if they seem more suitable for particular problems, and to use alternative methods when special features suggest them. But, for the present, 'partial fractions' means the form given in the text.

2. Preliminary treatment of simple examples. Before the general treatment, an elementary account is given of a way in which simple cases are tackled. (This may be sufficient at first.)

Consider, for instance, the expression

$$E \equiv \frac{3}{(1+x)(4+3x)};$$

the aim is to put it into the form

$$\frac{\text{something}}{1+x}+\frac{\text{something}}{4+3x},$$

say

$$\frac{a}{1+x}+\frac{b}{4+3x},$$

where a, b are two numbers independent of x. This requires the identity

$$\frac{3}{(1+x)(4+3x)} \equiv \frac{a}{1+x}+\frac{b}{4+3x}$$

$$\equiv \frac{a(4+3x)+b(1+x)}{(1+x)(4+3x)},$$

so that, the numerators being the same since the denominators are,

$$3 \equiv a(4+3x)+b(1+x),$$

an identity required to be true for all values of x. Put $x = -1$; then (since the term multiplying b is zero for $x = -1$)

$$a = \frac{3}{4+3(-1)} = 3.$$

Put $x = -\frac{4}{3}$; then (since the term multiplying a is zero for $x = -\frac{4}{3}$)

$$b = \frac{3}{1+(-\frac{4}{3})} = -9.$$

Thus, with these values for a and b,

$$E \equiv \frac{3}{1+x} - \frac{9}{4+3x},$$

provided, of course, that the form

$$\frac{a}{1+x} + \frac{b}{4+3x}$$

was a possible one in the first place. This proviso will be justified in the subsequent paragraphs; for the present, the accuracy of the result is easily verified by expressing the two terms on the right with a common denominator again, when the given fraction is recovered.

An examination of the working reveals an exceedingly simple rule for calculating the coefficients. Write a, b in their elementary forms first obtained:

$$a = \frac{3}{4+3(-1)}, \quad b = \frac{3}{1+(-\frac{4}{3})}.$$

Then the identity is

$$\frac{3}{(1+x)(4+3x)} \equiv \frac{3}{(1+x)[4+3(-1)]} + \frac{3}{[1+(-\frac{4}{3})](4+3x)}.$$

In other words, *the coefficient of* $1/(1+x)$ *is found by putting* -1 *for* x *in the rest of the expression*, and *the coefficient of* $1/(4+3x)$ *is found by putting* $-\frac{4}{3}$ *for* x *in the rest of the expression.*

This rule is sometimes called the *cover-up* rule; each coefficient is found by *covering up the relevant linear factor and replacing x in the rest by the result of equating the covered factor to zero.*

The two Illustrations which follow demonstrate the application of the rule. The method deals with all the simplest cases, but harder examples will require the more elaborate treatment given in the rest of the chapter. (The proof that the rule works generally is given in the next paragraph.)

***Illustration* 1.** *To express*

$$E \equiv \frac{17x+11}{(x+1)(x-2)(x+3)}$$

in the form

$$\frac{a}{x+1} + \frac{b}{x-2} + \frac{c}{x+3}.$$

(i) To find a, omit the factor $(x+1)$:

$$\frac{17x+11}{(x-2)(x+3)}.$$

Replace x by -1:

$$a = \frac{17(-1)+11}{(-1-2)(-1+3)} = \frac{-6}{(-3)(2)} = 1.$$

(ii) To find b, omit the factor $(x-2)$:

$$\frac{17x+11}{(x+1)(x+3)}.$$

Replace x by 2: $$b = \frac{17(2)+11}{(2+1)(2+3)} = \frac{45}{(3)(5)} = 3.$$

(iii) To find c, omit the factor $(x+3)$:

$$\frac{17x+11}{(x+1)(x-2)}.$$

Replace x by -3:

$$c = \frac{17(-3)+11}{(-3+1)(-3-2)} = \frac{-40}{(-2)(-5)} = -4.$$

Thus $$E \equiv \frac{1}{x+1} + \frac{3}{x-2} - \frac{4}{x+3}.$$

Illustration 2. *To express*

$$E \equiv \frac{5x^2+38x-91}{(x-1)(x+3)(x-5)}$$

in the form $$\frac{a}{x-1} + \frac{b}{x+3} + \frac{c}{x-5}.$$

$$a = \frac{5(1^2)+38(1)-91}{(1+3)(1-5)} = \frac{-48}{-16} = 3;$$

$$b = \frac{5(-3)^2+38(-3)-91}{(-3-1)(-3-5)} = \frac{-160}{32} = -5;$$

$$c = \frac{5(5)^2+38(5)-91}{(5-1)(5+3)} = \frac{224}{32} = 7.$$

Hence $$E \equiv \frac{3}{x-1} - \frac{5}{x+3} + \frac{7}{x-5}.$$

Examples 1

Apply the method used in the two preceding Illustrations to the following examples. *After acquiring reasonable fluency, return to work the examples mentally.*

1. $\dfrac{2}{(x+1)(x+2)}$.

2. $\dfrac{2x}{(x-1)(x-2)}$.

3. $\dfrac{x+1}{(x-1)(x+2)}$.

4. $\dfrac{2x+3}{(x+1)(x+2)}$.

5. $\dfrac{6}{(x+1)(x+2)(x+3)}$.

6. $\dfrac{6x}{(x-1)(x-2)(x-3)}$.

7. $\dfrac{6x+6}{(x-1)(x-2)(x-3)}$.

8. $\dfrac{6x^2}{(x+1)(x+2)(x+3)}$.

9. $\dfrac{x+4}{x(x+2)}$.

10. $\dfrac{x-4}{x(x-2)}$.

11. $\dfrac{x+1}{x(x-1)(x-2)}$.

12. $\dfrac{x^2+1}{x(x-1)(x+2)}$.

3. The general method. If $F(x)$, $G(x)$ are two *polynomials* in a variable x, the quotient

$$\frac{F(x)}{G(x)}$$

is called a *rational function* of x. The particular case $G(x) \equiv 1$ shows that polynomials are included among the rational functions. If the degree of $F(x)$ is greater than that of $G(x)$, then $F(x)$ can be divided by $G(x)$, giving a quotient $Q(x)$ and remainder $R(x)$. Thus

$$F(x) \equiv Q(x)G(x)+R(x),$$

where the degree of $R(x)$ is less than that of $G(x)$. It is assumed that, if necessary, this step has been taken; and *attention is consequently confined in the theory to rational functions $F(x)/G(x)$ for which the degree of $F(x)$ is less than that of $G(x)$.*

If $G(x)$ is of degree n, then (p. 55) $G(x)$ can be expressed as a product of n linear factors in the form

$$A(x-a)(x-b)\dots(x-k),$$

where $a, b, \dots, k$ may be complex, and need not be distinct.

Suppose, in particular, that $x-a$ is one of the factors of the denominator $G(x)$, having multiplicity p. Then $G(x)$ can be written in product form:

$$G(x) \equiv (x-a)^p g(x) \qquad g(a) \neq 0,$$

where $g(x)$ is a polynomial of degree $n-p$ not divisible by $x-a$. The numerator is a polynomial whose degree is, by assumption, less than n; it is agreed to call the degree $n-1$ on the understanding that some of the 'top' coefficients may be zero for any particular example. To emphasize its degree, the numerator may be denoted by the symbol

$$f_{n-1}(x),$$

and the rational function then written in the form

$$\frac{f_{n-1}(x)}{(x-a)^p g(x)}.$$

We define the *prime partial fraction* corresponding to the factor $x-a$ as the fraction, with denominator $(x-a)^p$, whose *numerator is formed by substituting a for x in the rest of the rational function*; that is, the numerator is the number

$$\frac{f_{n-1}(a)}{g(a)}.$$

(This is the coefficient obtained by the 'cover-up' rule (p. 106) for simple cases.) Thus the prime partial fraction is

$$\frac{f_{n-1}(a)}{(x-a)^p g(a)},$$

with a written for x everywhere except in the factor $(x-a)^p$.

***Illustration* 3.** In order to make the subsequent argument clearer, a numerical example is given to show what will be involved (compare also §2, pp. 105–107).

To find an expression for the function

$$\frac{x^2+x+6}{(x-1)^2(3x+1)}.$$

The prime partial fraction corresponding to the factor $x-1$ is

$$\frac{1^2+1+6}{(x-1)^2(3.1+1)} = \frac{2}{(x-1)^2}.$$

If this fraction is *subtracted from the given expression*, the result is

$$\frac{x^2+x+6}{(x-1)^2(3x+1)} - \frac{2}{(x-1)^2},$$

or, over a common denominator,

$$\frac{x^2+x+6-2(3x+1)}{(x-1)^2(3x+1)} = \frac{x^2-5x+4}{(x-1)^2(3x+1)}.$$

Now the point of the proof which follows is that the new numerator x^2-5x+4 *must* contain as a factor the term $x-1$ whose first prime partial fraction was subtracted. (It may, exceptionally, contain that term to a power higher than the first, but that cannot be foreseen.) Here, the numerator is

$$x^2-5x+4 \equiv (x-1)(x-4),$$

so that
$$\frac{x^2+x+6}{(x-1)^2(3x+1)}-\frac{2}{(x-1)^2}=\frac{(x-1)(x-4)}{(x-1)^2(3x+1)},$$

or
$$\frac{x^2+x+6}{(x-1)^2(3x+1)}=\frac{2}{(x-1)^2}+\frac{x-4}{(x-1)(3x+1)}.$$

Illustration 4. *To find similarly an expression for the function*

$$\frac{x^2+2x+10}{(x^2+4)(x+3)}.$$

The prime partial fraction corresponding to the linear factor $x+3$ is

$$\frac{(-3)^2+2(-3)+10}{(x+3)\{(-3)^2+4\}}=\frac{1}{x+3}.$$

If this fraction is subtracted from the given expression, the result is

$$\frac{x^2+2x+10}{(x^2+4)(x+3)}-\frac{1}{x+3},$$

or, over a common denominator,

$$\frac{x^2+2x+10-(x^2+4)}{(x^2+4)(x+3)}=\frac{2x+6}{(x^2+4)(x+3)}.$$

The numerator is $2(x+3)$, *having* $x+3$ *as a factor.*

Hence
$$\frac{x^2+2x+10}{(x^2+4)(x+3)}-\frac{1}{x+3}=\frac{2(x+3)}{(x^2+4)(x+3)}$$
$$=\frac{2}{x^2+4},$$

so that
$$\frac{x^2+2x+10}{(x^2+4)(x+3)}=\frac{1}{x+3}+\frac{2}{x^2+4}.$$

Compare also the later Illustrations, to be given (p. 113) after the general theory is established. [The beginner may prefer to omit the general theory now and to go straight to the Illustrations; but he should return to make quite sure of the structure of the argument.]

The basic theorem may now be proved:

The given rational function

$$\frac{f_{n-1}(x)}{(x-a)^p g(x)}$$

can be expressed as the sum of its prime partial fraction and a rational function of reduced denominator, in the form

$$\frac{f_{n-1}(x)}{(x-a)^p g(x)}\equiv\frac{f_{n-1}(a)}{(x-a)^p g(a)}+\frac{f_{n-2}(x)}{(x-a)^{p-1}g(x)},$$

where $f_{n-2}(x)$ *is a polynomial of degree* $n-2$ *(or less).*

Consider the difference of the function and the prime partial fraction:

$$\frac{f_{n-1}(x)}{(x-a)^p g(x)} - \frac{f_{n-1}(a)}{(x-a)^p g(a)} \equiv \frac{g(a)f_{n-1}(x) - f_{n-1}(a)g(x)}{(x-a)^p g(x)g(a)}.$$

This expression may be written more briefly as, say,

$$\frac{n(x)}{(x-a)^p g(x)g(a)},$$

where the numerator $n(x)$ is given by the identity

$$n(x) \equiv g(a)f_{n-1}(x) - f_{n-1}(a)g(x).$$

Since

$$\begin{aligned} n(a) &= g(a)f_{n-1}(a) - f_{n-1}(a)g(a) \\ &= 0, \end{aligned}$$

the polynomial $n(x)$, by the Remainder Theorem, has $x-a$ as a factor. Moreover, the degree of $n(x)$ cannot exceed the greater of $n-1$, of $f_{n-1}(x)$, and $n-p$, of $g(x)$, so that the degree of $n(x)$ may be taken as $n-1$ (with the possibility of zero 'top' coefficients as before). Hence $n(x)$ is of the form

$$n(x) \equiv (x-a)f_{n-2}(x)g(a),$$

where $f_{n-2}(x)$ is a certain polynomial of degree $n-2$; the numerical factor $g(a)$ is inserted merely to preserve symmetry of notation later —its value is not zero, as we have seen. Thus

$$\begin{aligned} \frac{f_{n-1}(x)}{(x-a)^p g(x)} - \frac{f_{n-1}(a)}{(x-a)^p g(a)} &\equiv \frac{(x-a)f_{n-2}(x)g(a)}{(x-a)^p g(x)g(a)} \\ &\equiv \frac{f_{n-2}(x)}{(x-a)^{p-1}g(x)}, \end{aligned}$$

on cancelling $g(a)$; and so

$$\frac{f_{n-1}(x)}{(x-a)^p g(x)} \equiv \frac{f_{n-1}(a)}{(x-a)^p g(a)} + \frac{f_{n-2}(x)}{(x-a)^{p-1}g(x)}.$$

The same reasoning may be applied to the rational function

$$\frac{f_{n-2}(x)}{(x-a)^{p-1}g(x)} \equiv \frac{f_{n-2}(a)}{(x-a)^{p-1}g(a)} + \frac{f_{n-3}(x)}{(x-a)^{p-2}g(x)},$$

and continued for diminishing powers of $x-a$ until that factor is extinguished. Thus

$$\frac{f_{n-1}(x)}{(x-a)^p g(x)} \equiv \frac{f_{n-1}(a)}{(x-a)^p g(a)} + \frac{f_{n-2}(a)}{(x-a)^{p-1}g(a)} + \ldots + \frac{f_{n-p}(a)}{(x-a)g(a)} + \frac{f_{n-p-1}(x)}{g(x)},$$

where the polynomials in the numerators are calculated successively by the method given in the proof of the basic theorem.

Illustration 5.* *To effect the reduction of the function*

$$\frac{6x^3-13x^2+18x+1}{(x-1)^3(x+2)}.$$

(i) The first prime partial fraction, corresponding to $x-1$, is

$$\frac{6.1^3-13.1^2+18.1+1}{(x-1)^3(1+2)},$$

or

$$\frac{4}{(x-1)^3},$$

and this is subtracted from the given function:

$$\frac{6x^3-13x^2+18x+1}{(x-1)^3(x+2)}-\frac{4}{(x-1)^3} \equiv \frac{6x^3-13x^2+18x+1-4x-8}{(x-1)^3(x+2)}$$

$$\equiv \frac{6x^3-13x^2+14x-7}{(x-1)^3(x+2)}.$$

The numerator must be divisible by $x-1$; the quotient may be calculated by the scheme:

$$\begin{array}{rrrr} 6 & -13 & 14 & -7 \\ & 6 & -7 & 7 \\ \hline 6 & -7 & 7 & \underline{|0} \end{array}$$

Thus $$\frac{6x^3-13x^2+18x+1}{(x-1)^3(x+2)}-\frac{4}{(x-1)^3} \equiv \frac{6x^2-7x+7}{(x-1)^2(x+2)}.$$

(ii) The new prime partial fraction is

$$\frac{6.1^2-7.1+7}{(x-1)^2(1+2)},$$

or

$$\frac{2}{(x-1)^2},$$

and this is subtracted from the difference just obtained:

$$\frac{6x^2-7x+7}{(x-1)^2(x+2)}-\frac{2}{(x-1)^2} \equiv \frac{6x^2-7x+7-2x-4}{(x-1)^2(x+2)} \equiv \frac{6x^2-9x+3}{(x-1)^2(x+2)}$$

$$\equiv \frac{3(x-1)(2x-1)}{(x-1)^2(x+2)} \equiv \frac{6x-3}{(x-1)(x+2)}.$$

Hence $$\frac{6x^3-13x^2+18x+1}{(x-1)^3(x+2)}-\frac{4}{(x-1)^3}-\frac{2}{(x-1)^2} \equiv \frac{6x-3}{(x-1)(x+2)}.$$

(iii) The final prime partial fraction is

$$\frac{6.1-3}{(x-1)(1+2)},$$

* This Illustration is intended to show how the preceding theory develops. It is not necessarily the quickest way of making the calculations for this particular problem. See p. 114.

or

$$\frac{1}{x-1},$$

and the last difference is

$$\frac{6x-3}{(x-1)(x+2)}-\frac{1}{x-1}\equiv\frac{6x-3-x-2}{(x-1)(x+2)}\equiv\frac{5(x-1)}{(x-1)(x+2)}\equiv\frac{5}{x+2}.$$

Hence $$\frac{6x^3-13x^2+18x+1}{(x-1)^3(x+2)}\equiv\frac{4}{(x-1)^3}+\frac{2}{(x-1)^2}+\frac{1}{x-1}+\frac{5}{x+2}.$$

The general discussion (from p. 111) may now be resumed.

When the factor $x-a$ has been eliminated from the denominator, the factor $x-b$ must next be considered. Suppose that $g(x)$ has $x-b$ as a factor of degree q, so that

$$g(x)\equiv(x-b)^q h(x),$$

where $x-b$ is not a factor of $h(x)$. Then the residual function

$$\frac{f_{n-p-1}(x)}{(x-b)^q h(x)}$$

can be treated in the same way as the original. This process can be repeated successively for all the factors in the denominator. Ultimately *the given expression is obtained in the form*

$$\frac{A_1}{(x-a)^p}+\ldots+\frac{A_p}{x-a}+\frac{B_1}{(x-b)^q}+\ldots+\frac{B_q}{x-b}+\frac{C_1}{(x-c)^r}+\ldots+\frac{C_r}{x-c}+\ldots,$$

where $x-a$, $x-b$, $x-c$, ... are the factors of the denominator, of multiplicities $p, q, r, \ldots$.

The given rational function is then said to be expressed in partial fractions.

The details of this exposition will become clearer after the particular examples of the following paragraph have been studied.

4. The calculations in practice. As has been said, the actual calculation of partial fractions may be carried out by many methods. It is probably true that no one method can be used exclusively to give the best results in all cases. We propose, however, to base the treatment almost entirely on the work of the preceding paragraph, with a certain amount of telescoping to shorten the number of steps. Any standard text-book on algebra gives alternative methods.

***Illustration* 6.** *To express the rational function*

$$\frac{x^3+7}{(x-1)(x-2)^2(x-3)^2}$$

in partial fractions.

The *first-power* factor $x-1$ can wait.* The process begins by reducing the two others simultaneously.

The prime partial fractions corresponding to 2, 3 are

$$\frac{2^3+7}{(2-1)(x-2)^2(2-3)^2}, \quad \frac{3^3+7}{(3-1)(3-2)^2(x-3)^2},$$

or

$$\frac{15}{(x-2)^2}, \quad \frac{17}{(x-3)^2},$$

so we consider the difference

$$\frac{x^3+7}{(x-1)(x-2)^2(x-3)^2}-\frac{15}{(x-2)^2}-\frac{17}{(x-3)^2}$$
$$\equiv \frac{x^3+7-15(x-1)(x-3)^2-17(x-1)(x-2)^2}{(x-1)(x-2)^2(x-3)^2}.$$

The numerator is

$$x^3+7-(x-1)\{15(x^2-6x+9)+17(x^2-4x+4)\}$$
$$\equiv x^3+7-(x-1)(32x^2-158x+203)$$
$$\equiv x^3+7-(32x^3-158x^2+203x)+(32x^2-158x+203)$$
$$\equiv -31x^3+190x^2-361x+210.$$

We know that $x-2$, $x-3$ are factors, so we find the quotient by synthetic division:

$$\begin{array}{rrrr}
-31 & 190 & -361 & 210 \\
 & -62 & 256 & -210 \\
\hline
-31 & 128 & -105 & \underline{0|} \\
 & -93 & 105 & \\
\hline
-31 & 35 & \underline{0|} &
\end{array}$$

Hence

$$\frac{x^3+7}{(x-1)(x-2)^2(x-3)^2}-\frac{15}{(x-2)^2}-\frac{17}{(x-3)^2} \equiv \frac{-31x+35}{(x-1)(x-2)(x-3)}.$$

The final step is immediate. The prime partial fractions of the right-hand side are

$$\frac{-31.1+35}{(x-1)(1-2)(1-3)}, \quad \frac{-31.2+35}{(2-1)(x-2)(2-3)}, \quad \frac{-31.3+35}{(3-1)(3-2)(x-3)},$$

or

$$\frac{2}{x-1}, \quad \frac{27}{x-2}, \quad \frac{-29}{x-3},$$

so that

$$\frac{-3x+35}{(x-1)(x-2)(x-3)} \equiv \frac{2}{x-1}+\frac{27}{x-2}-\frac{29}{x-3}.$$

Hence the given rational functon is

$$\frac{15}{(x-2)^2}+\frac{27}{x-2}+\frac{17}{(x-3)^2}-\frac{29}{x-3}+\frac{2}{x-1}.$$

* This detail of manipulation helps to keep calculations shorter.

REMARK. The numbers in the numerators imply that the calculations will be unpleasant by any method. Here the division by $x-2$, $x-3$ without remainder provides a useful check on accuracy.

Illustration 7. To express

$$E \equiv \frac{x^3+x+1}{(x-2)^4}$$

in partial fractions.

This example is inserted to remind the reader to look at the particular features of a special problem before applying the standard technique too blindly. Essentially, it is concerned only with $x-2$, so we write, temporarily,

$$x-2 = y.$$

Then
$$x^3+x+1 \equiv (y+2)^3+(y+2)+1$$
$$\equiv y^3+6y^2+13y+11,$$

so that
$$E \equiv \frac{y^3+6y^2+13y+11}{y^4}$$
$$= \frac{11}{y^4}+\frac{13}{y^3}+\frac{6}{y^2}+\frac{1}{y}$$
$$\equiv \frac{11}{(x-2)^4}+\frac{13}{(x-2)^3}+\frac{6}{(x-2)^2}+\frac{1}{x-2}.$$

More elaborate uses of this device will be given later (p. 213).

Examples 2

Express in partial fractions:

1. $\dfrac{1}{(x-1)^2(x-2)}$.

2. $\dfrac{x}{(x+1)^2(x-2)}$.

3. $\dfrac{x+1}{x^2(x-1)}$.

4. $\dfrac{x^2}{(x-1)^2(x-2)^2(x-3)}$.

5. $\dfrac{27x}{(x^2-1)(x+2)^3}$.

6. $\dfrac{2x+3}{(x+2)^2(x+1)}$.

7. $\dfrac{x^2}{(x+2)^2(x-1)}$.

8. $\dfrac{x^3}{(x-1)^3(x+1)}$.

9. $\dfrac{4}{(x-1)^3(x-2)^2(x-3)}$.

10. $\dfrac{x^4}{(x-1)^4(x+2)}$.

11. $\dfrac{x+2}{(x-1)^5}$.

12. $\left(\dfrac{x-1}{x+1}\right)^4$.

5. Quadratic factors. When the denominator $G(x)$ of the rational function

$$\frac{F(x)}{G(x)}$$

has complex zeros, so that the equation

$$G(x) = 0$$

has complex roots, the procedure already given can be used, but some change of form may then be thought desirable for the final answer. The coefficients in $G(x)$ are assumed real, so that (p. 56) complex roots occur in conjugate pairs. The coefficients in $F(x)$ are also assumed real.

Suppose that $u \pm iv$ are complex roots of the equation $G(x) = 0$, each occurring with multiplicity p. Thus, if $G(x)$ is of degree n,

$$G(x) \equiv \{(x-u)^2+v^2\}^p g(x),$$

where $g(x)$ is a polynomial, with real coefficients, of degree $n-2p$. Hence the rational function is

$$\frac{F(x)}{\{(x-u)^2+v^2\}^p g(x)}.$$

To conform with the notation used earlier, write the function in the form (with F replaced by f)

$$\frac{f(x)}{\{(x-u)^2+v^2\}^p g(x)}.$$

For convenience, denote the zeros of $(x-u)^2+v^2$, which are complex conjugates, by the notation $\alpha, \bar{\alpha}$, so that $\alpha = u+iv$, $\bar{\alpha} = u-iv$. Then the *prime partial fraction* corresponding to the factor $(x-u)^2+v^2$ is defined to be the function

$$\frac{Ax+B}{\{(x-u)^2+v^2\}^p},$$

where *the linear polynomial* $Ax+B$ *is that one which takes the value* $f(\alpha)/g(\alpha)$ *when* $x = \alpha$ *and* $f(\bar{\alpha})/g(\bar{\alpha})$ *when* $x = \bar{\alpha}$. Thus (p. 12)

$$Ax+B \equiv \left(\frac{x-\bar{\alpha}}{\alpha-\bar{\alpha}}\right)\frac{f(\alpha)}{g(\alpha)}+\left(\frac{x-\alpha}{\bar{\alpha}-\alpha}\right)\frac{f(\bar{\alpha})}{g(\bar{\alpha})},$$

though this explicit formulation is not always necessary. (See Illustration 8.)

Note that, in spite of their complex form, the coefficients A, B are actually real.

As in the earlier and simpler case (p. 109), the argument is interrupted to deal with a particular example.

Illustration 8. *To find an expression for the function*

$$\frac{6x-8}{(x^2+2x+5)(x^2-4x+13)}.$$

The prime partial fraction corresponding to the factor x^2+2x+5, whose zeros are $-1\pm 2i$, is

$$\frac{Ax+B}{x^2+2x+5},$$

where $Ax+B$ is the linear polynomial which takes the values

$$\frac{6(-1\pm 2i)-8}{(-1\pm 2i)^2-4(-1\pm 2i)+13}$$

when $x = -1\pm 2i$.

Taking the positive alternative sign only, the relation is

$$\begin{aligned} A(-1+2i)+B &= \frac{-14+12i}{(1-4i-4)-4(-1+2i)+13} \\ &= \frac{-14+12i}{14-12i} \\ &= -1. \end{aligned}$$

Since A, B are real, the real and imaginary parts of this relation can be equated, giving

$$-A+B = -1, \quad 2A = 0,$$

so that

$$A = 0, \quad B = -1.$$

Thus the prime partial fraction is

$$\frac{-1}{x^2+2x+5}.$$

Following the earlier procedure, consider the difference

$$\begin{aligned} \frac{6x-8}{(x^2+2x+5)(x^2-4x+13)}-\frac{-1}{x^2+2x+5} &\equiv \frac{6x-8+(x^2-4x+13)}{(x^2+2x+5)(x^2-4x+13)} \\ &\equiv \frac{x^2+2x+5}{(x^2+2x+5)(x^2-4x+13)}. \end{aligned}$$

The subsequent theory will prove that the term x^2+2x+5 must be a factor of the numerator; in this simple case the fact is obvious, since it *is* the numerator. Thus

$$\frac{6x-8}{(x^2+2x+5)(x^2-4x+13)}+\frac{1}{x^2+2x+5} \equiv \frac{1}{x^2-4x+13},$$

so that

$$\frac{6x-8}{(x^2+2x+5)(x^2-4x+13)} \equiv \frac{1}{x^2-4x+13}-\frac{1}{x^2+2x+5}.$$

Returning to the general discussion (p. 116) consider the difference between the given expression and the prime partial fraction:

$$\frac{f(x)}{\{(x-u)^2+v^2\}^p g(x)}-\frac{Ax+B}{\{(x-u)^2+v^2\}^p} \equiv \frac{f(x)-(Ax+B)g(x)}{\{(x-u)^2+v^2\}^p g(x)}.$$

Write this expression in the form, say,

$$\frac{n(x)}{\{(x-u)^2+v^2\}^p g(x)},$$

where the numerator $n(x)$ is given by the identity

$$n(x) \equiv f(x)-(Ax+B)g(x).$$

If $\alpha \equiv u+iv$, $\bar{\alpha} \equiv u-iv$ are the zeros of $(x-u)^2+v^2$, substitution of α for x gives

$$n(\alpha) = f(\alpha)-(A\alpha+B)g(\alpha).$$

But, by definition of the polynomial $Ax+B$ (p. 116),

$$A\alpha+B = \frac{f(\alpha)}{g(\alpha)},$$

so that

$$n(\alpha) = f(\alpha)-\frac{f(\alpha)}{g(\alpha)}g(\alpha)$$
$$= 0.$$

Thus, by the Remainder Theorem, $x-\alpha$ is a factor of $n(x)$; similarly, $x-\bar{\alpha}$ is also a factor of $n(x)$. Hence $n(x)$ has as a factor the product

$$(x-\alpha)(x-\bar{\alpha}) \equiv \{(x-u)^2+v^2\};$$

say

$$n(x) \equiv \{(x-u)^2+v^2\}m(x).$$

Thus

$$\frac{f(x)}{\{(x-u)^2+v^2\}^p g(x)} \equiv \frac{Ax+B}{\{(x-u)^2+v^2\}^p}+\frac{m(x)}{\{(x-u)^2+v^2\}^{p-1}g(x)}.$$

The whole process may now be repeated for the new expression

$$\frac{m(x)}{\{(x-u)^2+v^2\}^{p-1}g(x)},$$

obtaining for it the form

$$\frac{A'x+B'}{\{(x-u)^2+v^2\}^{p-1}}+\frac{q(x)}{\{(x-u)^2+v^2\}^{p-2}g(x)}.$$

By successive stages, the quadratic factor $(x-u)^2+v^2$ is thus eliminated from the denominator. As a result, the given function is expressed in the form

$$\frac{Ax+B}{\{(x-u)^2+v^2\}^p}+\frac{A'x+B'}{\{(x-u)^2+v^2\}^{p-1}}+\ldots+\frac{A'''x+B'''}{(x-u)^2+v^2}+\frac{r(x)}{g(x)}.$$

Attention may now be given to the factors of $g(x)$ and the argument repeated as often as is necessary.

Examples 3

Express in partial fractions:

1. $\dfrac{1}{(x^2+1)^2(x-1)}$.

2. $\dfrac{x+1}{(x^2+1)^2(x^2+2)}$.

3. $\dfrac{x^2+1}{(x^2+x+1)(x^2-x+1)}$.

4. $\dfrac{x}{(x^2+x+1)(x^2-x+1)}$.

5. $\dfrac{(x+1)^2}{(x^2+x+1)(x^2-x+1)}$.

6. $\dfrac{9}{(x^2+4)^2(x^2+1)}$.

7. $\dfrac{1}{(x^2+1)(x+1)^2(x-1)}$.

8. $\dfrac{x^2+1}{(x^2+3)^3(x-1)}$.

6. Uniqueness of the form. To maintain continuity of argument, we have slurred over one point which must be examined explicitly. The method adopted for partial fractions has been to calculate the successive coefficients corresponding to the various factors of the denominator, but it is not immediately obvious that the answers obtained do not depend on the order in which those factors are selected. The point is, however, easily settled. Suppose that, for a factor $x-a$ occurring to power n, two different orders of calculation give the contributions

$$\frac{A_1}{(x-a)^n}+\frac{A_2}{(x-a)^{n-1}}+\ldots+\frac{A_n}{x-a},$$

$$\frac{B_1}{(x-a)^n}+\frac{B_2}{(x-a)^{n-1}}+\ldots+\frac{B_n}{x-a},$$

and that the parts independent of $x-a$ are then $\phi(x)$, $\psi(x)$ respectively. Thus

$$\frac{A_1}{(x-a)^n}+\frac{A_2}{(x-a)^{n-1}}+\ldots+\frac{A_n}{x-a}+\phi(x)$$

$$\equiv\frac{B_1}{(x-a)^n}+\frac{B_2}{(x-a)^{n-1}}+\ldots+\frac{B_n}{x-a}+\psi(x),$$

each being identical to the given expression.

If $A_1 = B_1$, omit that term from each side; if $A_2 = B_2$, omit that term from each side; let k be the first number such that $A_k \neq B_k$. Then

$$\frac{A_k}{(x-a)^{n-k+1}}+\ldots+\frac{A_n}{x-a}+\phi(x) \equiv \frac{B_k}{(x-a)^{n-k+1}}+\ldots+\frac{B_n}{x-a}+\psi(x).$$

Multiply throughout by $(x-a)^{n-k+1}h(x)$, where $h(x)$ is the least common multiple of the denominators of $\phi(x)$, $\psi(x)$:

$$h(x)\{A_k+A_{k+1}(x-a)+\ldots+A_n(x-a)^{n-k}+(x-a)^{n-k+1}\phi(x)\}$$
$$\equiv h(x)\{B_k+B_{k+1}(x-a)+\ldots+B_n(x-a)^{n-k}+(x-a)^{n-k+1}\psi(x)\}.$$

Now both sides of this identity are polynomials, and they are, in particular, equal when $x = a$, giving

$$A_k h(a) = B_k h(a).$$

Also $x-a$ is not a factor of the denominators of $\phi(x)$, $\psi(x)$, so that $h(a) \neq 0$. Hence

$$A_k = B_k.$$

The hypothesis that there exists a k such that $A_k \neq B_k$ is therefore untenable, so that the two forms of expression are in fact identical.

Similar reasoning may be applied also to quadratic factors

$$(x-u)^2+v^2.$$

Revision Examples V

1. Evaluate the constants on the right-hand side of the identity

$$\frac{x^4+1}{x^2-x-2} \equiv x^2+ax+b+\frac{c}{x-2}+\frac{d}{x+1}.$$

Resolve
$$\frac{x^4+1}{x^3+x^2+x+1}$$
into partial fractions and such other terms as may be necessary.

2. Find the values of a, b, c if

$$\frac{x^3-1}{(x+1)(x+2)} \equiv x+a+\frac{b}{x+1}+\frac{c}{x+2}.$$

Solve the equation

$$\frac{x^3-1}{(x+1)(x+2)} = x-3+\frac{1}{x+1}+\frac{1}{x+2}.$$

3. Express as the sum of partial fractions

(i) $\dfrac{x}{(x-1)(x-2)}$, (ii) $\dfrac{x^2}{(x-1)^2(x-2)}$, (iii) $\dfrac{x^3}{(x-1)^3}$.

4. Express in partial fractions

$$\frac{1}{x^3-1},\quad \frac{1}{x^3+1}.$$

Hence, or otherwise, find partial fractions for

$$\frac{1}{x^6-1},\quad \frac{x^3}{x^6-1}.$$

Find also the partial fractions for

$$\frac{x}{x^6-1},\quad \frac{x^2}{x^6-1}.$$

5. Express in partial fractions

$$\frac{1}{1-x-x^3+x^4}.$$

6. Express $\dfrac{1}{x^4+1},\quad \dfrac{1}{x^8-1}$

in partial fractions with real quadratic denominators.

[*Hint*: Consider $(x^2+1)^2-2x^2$.]

7. Express in partial fractions

$$\frac{1}{x(x+1)^2(x-1)^3},\quad \frac{1}{x(1+x^2)^2}.$$

8. Express in partial fractions

$$\frac{2x^5}{(x^2-1)(x^2-4)},\quad \frac{1}{x^8+x^7-x^4-x^3},$$

$$\frac{1+x^3}{1-x^3},\quad \frac{1}{(1-x)^2(1+x^2)}.$$

Warning Examples

1. Criticize the following 'argument':

Suppose that $\dfrac{x}{(x-1)\surd(x^2+1)} \equiv \dfrac{A}{x-1}+\dfrac{Bx+C}{\surd(x^2+1)}.$

This requires $\quad x \equiv A\surd(x^2+1)+(Bx+C)(x-1).$

Put $x = 0$. Then $\quad 0 = A-C.$

Put $x = 1$. Then $\quad 1 = A\surd 2.$

Put $x = -1$. Then $\quad -1 = A\surd 2-2(-B+C).$

Hence $\quad A = 1/\surd 2,\quad C = 1/\surd 2,\quad B = (1-\surd 2)/\surd 2,$

so that $\quad \dfrac{x}{(x-1)\surd(x^2+1)} \equiv \dfrac{1}{(x-1)\surd 2}+\dfrac{(1-\surd 2)x+1}{\surd 2\surd(x^2+1)}.$

In particular, let $x = 7$, so that $\surd(x^2+1) = 5\surd 2$. Then

$$\frac{7}{6.5\surd 2} = \frac{1}{6\surd 2} + \frac{(1-\surd 2)7+1}{10},$$

or

$$\frac{1}{15\surd 2} = \frac{8-7\surd 2}{10},$$

so that

$$2 = 3\surd 2(8-7\surd 2),$$

or

$$6\surd 2 = 11.$$

2. Criticize the following 'argument':

Suppose that $$\frac{1}{(x-1)\sin\frac{1}{2}\pi x} \equiv \frac{A}{x-1} + \frac{B}{\sin\frac{1}{2}\pi x}.$$

This requires $$1 \equiv A\sin\tfrac{1}{2}\pi x + B(x-1).$$

Put $x = 0$, Then $$1 = -B.$$

Put $x = 1$. Then $$1 = A.$$

Hence $$\frac{1}{(x-1)\sin\frac{1}{2}\pi x} = \frac{1}{x-1} - \frac{1}{\sin\frac{1}{2}\pi x}.$$

In particular, let $x = 5$, so that $\sin\frac{1}{2}\pi x = \sin\frac{5}{2}\pi = 1$. Then $\frac{1}{4} = \frac{1}{4} - 1$.

11

INEQUALITIES

The polynomial $$x^3-6x^2+11x-6$$

is very seldom equal to zero; indeed, it is zero only for the three values 1, 2, 3 of x. Otherwise it is either positive or negative.

The work of this chapter is intended to clarify the conditions for inequality.

1. The algebra of inequalities. The symbol $>$ is used (p. 4) to denote the phrase 'greater than', and the symbol $<$ to denote 'less than'. Great care should be taken in the use of these symbols, and under no circumstances should they be manipulated in the manner familiar for equalities, except after close examination.

It is assumed that all numbers are *real.*

Certain rules are self-evident, and may be stated without comment:

(i) If $$a > b, \quad b > c,$$
then $$a > c.$$

(ii) If $$a < b, \quad b < c,$$
then $$a < c.$$

(iii) If $$a > b,$$
then $$a+k > b+k.$$

(iv) If $$a > b, \quad c > d,$$
then $$a+c > b+d.$$

(v) If $$a < b, \quad c < d,$$
then $$a+c < b+d.$$

NOTE. If $a > b$, $c < d$, then no comparison can be made between $a+c$ and $b+d$. For example, $2 > 1$, $3 < 5$ and $2+3 < 1+5$; but $6 > 2$, $3 < 4$ and $6+3 > 2+4$.

The 'multiplication' of inequalities is a common source of trouble. The safest rule is *Never multiply or divide inequalities except by numbers known to be positive.* Whenever there is uncertainty about sign, multiplication or division should be approached warily.

Warning Illustrations

It might be thought that, if

$$a > b, \quad c > d,$$

then $ac > bd$. But consider the inequalities

$$3 > 2, \quad -4 > -5.$$

It is not true that $$3(-4) > 2(-5),$$

or $$-12 > -10.$$

Again, it might be thought that, if

$$a > b,$$

then $ka > kb$. But consider the inequality

$$3 > 2$$

and take $k = -5$. It is not true that

$$-15 > -10.$$

Note. The author has often seen the 'argument'

$$x^2 > 4,$$

therefore $$x > \pm 2.$$

This is complete nonsense, and quite meaningless. The correct deduction is that, if

$$x^2 > 4,$$

then *either* $x > +2$ *or* $x < -2$.

With the 'warning illustrations' in mind, we continue to formulate the rules:

(vi) If $$a > b$$

and if k is positive, then $$ka > kb.$$

(vii) If a, b, c, d are all positive, and if

$$a > b, \quad c > d,$$

then $$ac > bd.$$

REMARK. Rules can be given for the multiplication of inequalities by negative numbers. It is possible to argue, however, that an excess of rules leads to confusion. Once the 'positive' rules are fully absorbed, the modifications will be grasped easily.

For example, if

$$a > b,$$

then $$5a > 5b,$$

so that $$0 > 5b-5a,$$
or $$-5b > -5a,$$
or, what is equivalent, $$-5a < -5b,$$

so that *multiplication by -5 has reversed the sense of the inequality.*

2. The 'perfect square' technique. Many inequalities are based on the familiar fact that the square of a real number is positive. For example, if a, b, c are three unequal numbers, then

$$(b-c)^2, \quad (c-a)^2, \quad (a-b)^2$$

are all positive, so that

$$(b-c)^2+(c-a)^2+(a-b)^2 > 0,$$

or $$2(a^2+b^2+c^2)-2(bc+ca+ab) > 0.$$

Divide* by the positive number 2, and add $bc+ca+ab$ to each side; this gives a standard inequality

$$a^2+b^2+c^2 > bc+ca+ab.$$

An important use of squares leads to the theorem of the arithmetic and geometric means:

If a, b are two unequal positive numbers, then their 'arithmetic mean' $\frac{1}{2}(a+b)$ exceeds their 'geometric mean' $\sqrt{(ab)}$. (Compare p. 188.)

The inequality $$(\sqrt{a}-\sqrt{b})^2 > 0$$

gives $$a-2\sqrt{(ab)}+b > 0,$$

or $$a+b > 2\sqrt{(ab)},$$

or $$\tfrac{1}{2}(a+b) > \sqrt{(ab)}.$$

The case of equality arises only when $(\sqrt{a}-\sqrt{b})^2 = 0$, that is, when $a = b$.

Examples 1

1. Write down all the positive integers which satisfy the two inequalities

$$x < 5, \quad xy < 10.$$

What is the corresponding result if

$$x \leqslant 5, \quad xy \leqslant 10?$$

* We shall usually omit explicit reference to such elementary steps.

2. Write down all the integers (positive or negative) which lie between the limits

(i) $1-\sqrt{10}$, $1+\sqrt{10}$;
(ii) $\sqrt{17}-2$, $\sqrt{17}+2$;
(iii) $\sqrt{7}+\sqrt{5}$, $\sqrt{7}-\sqrt{5}$.

3. Find the least value, for varying x, of each of the functions:

(i) $(x-2)^2+3$. (ii) $(x-2)^2-3$.
(iii) $(x+1)^2+5$. (iv) $(x-5)^2+1$.
(v) x^2-2x+2. (vi) x^2-4x+7.
(vii) $x^2-6x+11$. (viii) x^2+4x+4.
(ix) $x^2+2x+17$. (x) $x^2-10x+30$.
(xi) x^4+1. (xii) x^4-2x^2+8.

3. Polynomial inequalities; graphical arguments. As a typical example, consider those values of x for which the polynomial

$$(x-1)(x-2)(x-3)$$

is positive, so that $(x-1)(x-2)(x-3) > 0$.

Attention must be given to the four intervals whose end-points are the zeros of the polynomial; that is, to the intervals

$$x < 1, \quad 1 < x < 2, \quad 2 < x < 3, \quad 3 < x.$$

A symbol such as $(+, -, +)$ will be used to denote that $x-1$ is positive, $x-2$ negative, $x-3$ positive. For the four intervals in turn, the symbols are

(i) $x < 1$, signs $(-, -, -)$, so that the product is negative;
(ii) $1 < x < 2$, signs $(+, -, -)$, so that the product is positive;
(iii) $2 < x < 3$, signs $(+, +, -)$, so that the product is negative;
(iv) $3 < x$, signs $(+, +, +)$, so that the product is positive.

Hence $$(x-1)(x-2)(x-3) > 0$$

when *either* $$1 < x < 2 \quad \textit{or} \quad 3 < x.$$

ALITER. A sketch of the graph

$$y = (x-1)(x-2)(x-3)$$

is shown in the diagram (fig. 11). It is at once evident that y is positive when x is between 1, 2 and also when x is greater than 3.

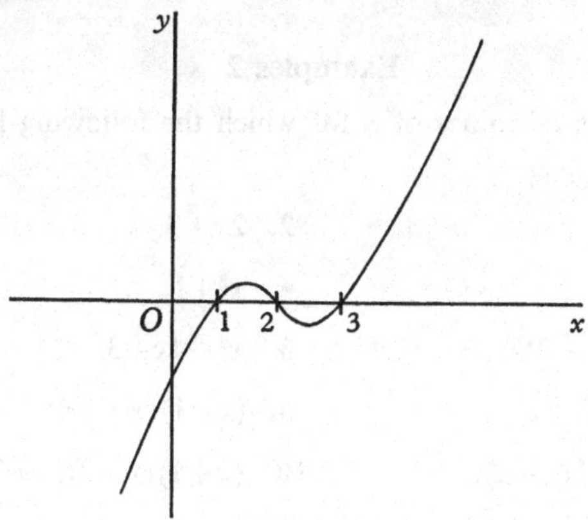

Fig. 11

Illustration **1.** *To find the values of x for which*

$$(x-1)^2(x-2) > 0.$$

Since $(x-1)^2$ is always positive, divide by it; the condition is thus

$$x-2 > 0,$$

so that

$$x > 2.$$

Illustration **2.** *To find the values of x for which*

$$x^3(x+2) > 0.$$

Since x^2 is always positive, divide by it; the condition is thus

$$x(x+2) > 0.$$

There are three intervals to be considered, $x < -2$, $-2 < x < 0$, $0 < x$.

(i) If $x < -2$, the signs are $(-, -)$, so that $x(x+2) > 0$;
(ii) If $-2 < x < 0$, the signs are $(-, +)$, so that $x(x+2) < 0$;
(iii) If $0 < x$, the signs are $(+, +)$, so that $x(x+2) > 0$.

Hence *either* $x < -2$ *or* $0 < x$.

The graph $y = x(x+2)$ is shown in the diagram (fig. 12).

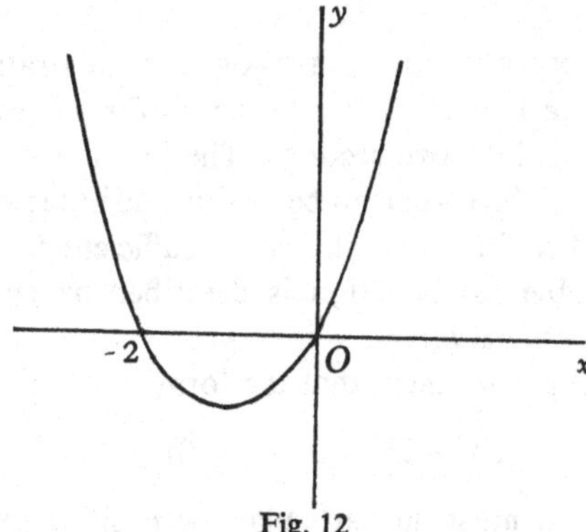

Fig. 12

Examples 2

State the ranges of values of x for which the following functions are positive:

1. $x-2$. **2.** $2x+3$.

3. x^3. **4.** x^2+1.

5. $(x-1)(x-2)$. **6.** x^2-4x+3.

7. $(x-1)^2(x+2)$. **8.** $(x-1)(x+2)^2$.

9. $x^3(x-1)^2(x-2)$. **10.** $(x+1)(x+2)(x+3)$.

11. x^2-4. **12.** x^3-8.

13. $(x-1)(x+2)(x-3)$. **14.** $(x+1)(x-2)(x+3)$.

15. $(x-1)(x-2)(x-3)(x-4)$.

16. $(x-1)^4(x-2)^3(x-3)^2(x-4)$.

4. The quadratic form. The quadratic polynomial

$$f(x) \equiv ax^2+2bx+c \quad (a \neq 0)$$

is also called a *quadratic form*.

Two points should be noticed. First, it is assumed that a is not zero, to ensure that the form is genuinely quadratic. Secondly, the number 2 is inserted before b for the convenience of the algebra which is to follow; this is our first use of such *binomial coefficients*, as they are called.

The problem is *to examine how the sign of $f(x)$ varies for varying values of x.*

(i) THE POSITIVE DEFINITE QUADRATIC FORM. A quadratic form is said to be *positive definite* if it is *greater then zero for all real values of x.*

The discussion falls into two sections. The first is to find conditions which are necessary if the form is to be positive definite; and the second is to examine whether these conditions are sufficient to ensure that it must be. This double investigation is described as giving a set of *necessary and sufficient conditions.*

(*a*) *Necessary.* Suppose, then, that the form

$$ax^2+2bx+c \quad (a \neq 0)$$

is positive definite. It must, in particular, be positive for large values of x, when (p. 93) the behaviour of the function is like that of the first term. Thus

$$ax^2\left(1+\frac{2b}{x}+\frac{c}{x^2}\right)$$

must be positive; and, for large x, the effect of the terms $2b/x$, c/x^2 is negligible, so that one necessary condition is

$$ax^2 > 0.$$

But $x^2 > 0$ always when x is real. Hence a *necessary* condition is

$$a > 0.$$

If
$$ax^2+2bx+c > 0,$$

both sides can be multiplied* throughout by a, *since it is positive*, without disturbing the inequality. Hence it is necessary that

$$a^2x^2+2abx+ac > 0,$$

or
$$a^2x^2+2abx+b^2+ac-b^2 > 0,$$

or
$$(ax+b)^2+ac-b^2 > 0.$$

Since the left-hand side is to be positive for all values of x, it must be so, in particular, for that value which makes it least; and that value is $x = -b/a$, as the term $(ax+b)^2$, being a square, is otherwise greater than zero. Hence it is necessary that

$$ac-b^2 > 0.$$

Thus *necessary conditions for the form*

$$ax^2+2bx+c$$

to be positive definite are

$$a > 0, \quad ac-b^2 > 0.$$

(*b*) *Sufficient*. We prove next that these conditions are sufficient, so that, *if* $a > 0$, $ac-b^2 > 0$, *then the form is positive definite.*

Observe first the identity

$$a(ax^2+2bx+c) \equiv (ax+b)^2+(ac-b^2).$$

If $ac-b^2$ is given to be positive, the right-hand side is positive, and so

$$a(ax^2+2bx+c) > 0.$$

But a is also given to be positive. Hence

$$ax^2+2bx+c > 0$$

for all values of x.

* It seems to be a general rule that whereas in numerical work we divide by the coefficient, in algebraic symbols it is more convenient to multiply.

NOTE. Since
$$ac > b^2,$$
it follows that ac is positive, so that (a being positive)
$$c > 0.$$
This condition could have been seen necessary at once by taking the particular value $x = 0$; we have preferred to use the alternative of $a > 0$ since this is, in fact, the first case of a much more general result.

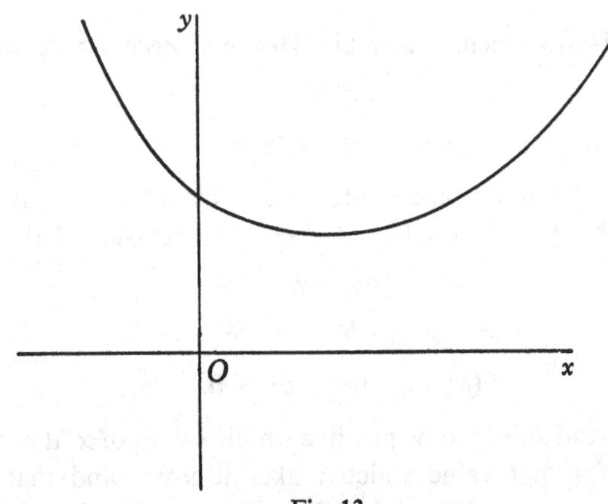

Fig. 13

The conditions may be illustrated graphically. If $ax^2+2bx+c > 0$, then the curve
$$y = ax^2+2bx+c$$
lies entirely 'above' the x-axis (fig. 13). Consideration of large values of x gives the condition $a > 0$ as before. Further, since the curve does not cut the x-axis, the roots of the equation
$$ax^2+2bx+c = 0$$
cannot be real. Hence (p. 20; but note the change in notation)
$$4b^2-4ac < 0,$$
so that
$$ac-b^2 > 0.$$

(ii) THE NEGATIVE DEFINITE QUADRATIC FORM. If the form
$$ax^2+2bx+c$$
is always negative, then the form
$$(-a)x^2+2(-b)x+(-c)$$
is always positive. Hence, as above,
$$(-a) > 0,$$
$$(-a)(-c)-(-b)^2 > 0,$$

so that
$$a < 0,$$
$$ac-b^2 > 0.$$

Thus the first inequality is reversed, but *the second is unchanged.*

(iii) QUADRATICS OF VARYING SIGN. The results of (i), (ii) together cover all the possibilities when the equation $ax^2+2bx+c = 0$ does not have real roots. We come now to the case
$$ac-b^2 < 0,$$
when the equation has real roots, say p, q. Then
$$ax^2+2bx+c \equiv a(x-p)(x-q).$$

If $p = q$, the form is positive definite ($a > 0$) or negative definite ($a < 0$) save that, instead of being always greater than (less than) zero, there is the one case, namely $x = p$, where it is actually equal to zero.

Suppose, for ease of statement, that
$$p > q.$$

Then
$$f(x) \equiv a(x-p)(x-q) \qquad (p > q).$$

Consider the first product
$$(x-p)(x-q).$$

For $x < q$, the signs are $(-, -)$, so that $(x-p)(x-q) > 0$; for $q < x < p$, the signs are $(-, +)$, so that $(x-p)(x-q) < 0$; for $p < x$, the signs are $(+, +)$, so that $(x-p)(x-q) > 0$.

Hence, *if a is positive* $f(x)$ is negative when x lies between p, q, but is otherwise positive.

If a is negative $f(x)$ is positive when x lies between p, q, but otherwise is negative.

The two cases are illustrated in the diagrams (fig. 14).

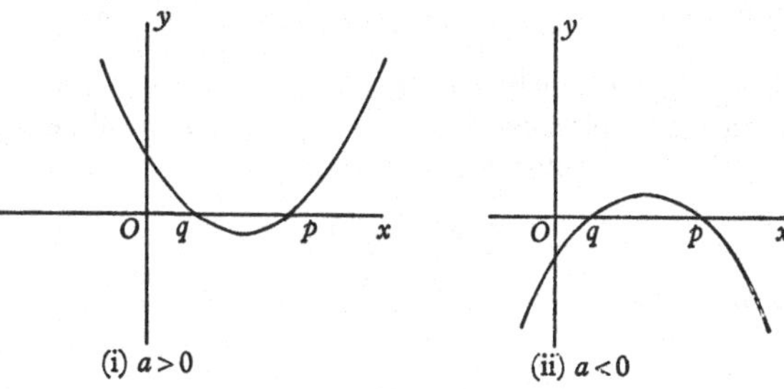

Fig. 14

Illustration 3. *To prove that, whatever the value of k, the quadratic form*

$$x^2+2(2k+1)x+6k^2+6k+4$$

is positive definite.

The coefficient of x^2 is unity, which is positive.

The condition $ac-b^2 > 0$ requires

$$(6k^2+6k+4)-(2k+1)^2 > 0,$$

or

$$2k^2+2k+3 > 0.$$

Now the form $2k^2+2k+3$ is itself positive definite, since

$$2 > 0,$$

and

$$2.3-1^2 > 0.$$

Hence

$$2k^2+2k+3 > 0,$$

and so, whatever the value of k, the given quadratic form is positive definite.

[Alternatively,

$$2k^2+2k+3 = 2(k^2+k+\tfrac{3}{2})$$

$$= 2\{(k+\tfrac{1}{2})^2+\tfrac{5}{4}\}$$

$$> 0.]$$

Illustration 4. *Cauchy's inequality. To prove that, if* $a_1, a_2, \ldots, a_n; b_1, b_2, \ldots, b_n$ *are real, then*

$$(a_1^2+a_2^2+\ldots+a_n^2)(b_1^2+b_2^2+\ldots+b_n^2) \geqslant (a_1b_1+a_2b_2+\ldots+a_nb_n)^2.$$

The expression

$$\sum_{r=1}^{n}(a_rx+b_r)^2$$

is greater than or equal to zero, since this is true for each individual term in the summation. Expressed as a quadratic in x, this gives the inequality

$$(\Sigma a_r^2)x^2+2(\Sigma a_rb_r)x+(\Sigma b_r^2) \geqslant 0,$$

and the condition for the left-hand side to be positive is (p. 129) precisely the required inequality

$$(\Sigma a_r^2)(\Sigma b_r^2) \geqslant (\Sigma a_rb_r)^2.$$

The case of equality arises when the quadratic in x has equal roots. But the quadratic, being a sum of squares, is zero only if the constituent terms are zero; that is, if there is a number x such that

$$a_rx+b_r = 0$$

for all values of r. Hence *the condition for equality is*

$$\frac{a_1}{b_1} = \frac{a_2}{b_2} = \ldots = \frac{a_n}{b_n}.$$

Examples 3

Find the conditions (if any) satisfied by k in order that the following functions may be positive definite:

1. x^2-2x+k.
2. x^2+4x+k.
3. kx^2-2x+2.
4. $x^2-2kx+1$.
5. x^2+6x+k.
6. kx^2+6x+9.
7. $x^2+2kx+10$.
8. x^2+8x+k.
9. $x^2+(k-1)(k-2)$.
10. x^2+k^2-4k+5.

Revision Examples VI

1. Prove that, if a, b are real,

$$ab \leqslant \left(\frac{a+b}{2}\right)^2,$$

and deduce that, if a, b, c, d are positive,

$$abcd \leqslant \left(\frac{a+b+c+d}{4}\right)^4.$$

By giving d a suitable value in terms of a, b, c, prove that, if a, b, c are positive,

$$abc \leqslant \left(\frac{a+b+c}{3}\right)^3.$$

2. If a, b, c, x, y, z are all real numbers, and

$$b+c \geqslant a, \quad c+a \geqslant b, \quad a+b \geqslant c,$$

show that the expression

$$a^2(x-y)(x-z)+b^2(y-x)(y-z)+c^2(z-x)(z-y)$$

is never negative.

3. If $\quad 0 < x < pr, \quad 0 < y < pr, \quad 0 < xy < p^2,$

show that $\quad x+y < (r+1/r)p.$

4. Establish necessary and sufficient conditions for $ax^2+2bx+c$ to be positive for all real values of x.

If all numbers involved are real, such that

$$a_1 > 0, \quad a_2 > 0, \quad a_3 > 0,$$

$$a_1c_1-b_1^2 > 0, \quad a_2c_2-b_2^2 > 0, \quad a_3c_3-b_3^2 > 0,$$

prove that

$$(a_1+a_2+a_3)(c_1+c_2+c_3)-(b_1+b_2+b_3)^2 > 0.$$

5. Prove that the inequalities

$$3x+2y > 6, \quad x-2y > 2, \quad -2x+y > 2$$

cannot be satisfied simultaneously.

6. Find for what values of a the inequalities

$$0 < x+2y < a, \quad xy > 3$$

can be satisfied simultaneously.

7. If a, b, c are positive, find conditions such that

$$(a-c)x^2+2(b-c)^2x+(a-c)^3$$

should be positive for all real values of x.

8. If x, y are real numbers connected by the relation

$$x^2+xy-2y^2-3x+3y+9 = 0,$$

prove that x may have any value, but that y cannot lie between 3 and -1.

9. Trace the changes of sign of the expression $2-x-x^2$ as x varies from $-\infty$ to $+\infty$ by marking segments of the x-axis with plus and minus signs, indicating clearly the values of x at the end points of the segments.

Trace in a similar manner the changes of sign of the expression $(x^2-3x+2)/(2x+1)$.

10. If a, b are positive numbers, prove that

$$(1-a)(1-b) > 1-a-b.$$

Deduce that, if a, b, c are positive, and one at least of a, b, c is less than unity, then

$$(1-a)(1-b)(1-c) > 1-a-b-c.$$

11. Factorize the expression

$$x^3+8x^2+7x,$$

and deduce the ranges of values of x for which it is positive.

12. Find the ranges of values of λ for which the roots of the equation

$$x^2+4x+3+\lambda(x^2-4x+3) = 0$$

are real.

13. Prove that the expression

$$(a+c)(ax^2+2bx+c)+2(b^2-ac)(x^2+1)$$

has the same sign for all values of x, provided that the equation $ax^2+2bx+c = 0$ has real roots.

14. Show that, if a, b, c are positive and $a+b \geqslant c$, then

$$\frac{a}{1+a}+\frac{b}{1+b} > \frac{c}{1+c}.$$

15. Prove that, if a, b are positive and x is positive or negative, then

$$(ab+x^2)(a+b) \geqslant 4abx.$$

16. Prove that

$$x^2(x+1)^2-2bx(x^2-1)+(b^2-2)(x-1)^2$$

is positive for all real values of x provided that b lies between $3-\sqrt{2}$ and $3+\sqrt{2}$.

17. If x, y, z are real numbers such that

$$x+y+z = 15, \quad yz+zx+xy = 72,$$

prove that

$$3 \leqslant x \leqslant 7.$$

18. If k is the least of $(b-c)^2, (c-a)^2, (a-b)^2$, where a, b, c are real, prove that

$$k \leqslant \tfrac{1}{2}(a^2+b^2+c^2).$$

19. If the sum of any two positive numbers a, b, c is greater than the third, prove that the same is true of the positive numbers

$$\sqrt{\{a(b+c-a)\}}, \quad \sqrt{\{b(c+a-b)\}}, \quad \sqrt{\{c(a+b-c)\}}.$$

20. The real numbers α, β, γ are not equal to -1 or $+1$, and are such that

$$\alpha^2+\beta^2+\gamma^2+2\alpha\beta\gamma = 1.$$

Prove that either all three lie between -1 and $+1$, or all three lie outside this range.

21. Find the ranges of values of x for which the functions

(i) $(x-1)^2(x-2)$,

(ii) $(x-1)(x-2)(x-3)$

are positive.

22. Prove that the function

$$x^2-8x+25$$

is always positive.

23. Find the range of values of λ for which the expression

$$x^2-4x+2+\lambda(x-4)^2$$

is positive for all real values of x.

24. Prove that the expression

$$x^2+2xy+4y^2+2x-10y+15$$

is positive for all real values of x and y, and that its least value is 2.

25. The polynomial $ax^2+2bx+c$ is positive for all real values of x. Find, in terms of a, b, c, k the greatest value that h can have if

$$ax^2+2bx+c \geqslant h(x-k)^2$$

for all real values of x.

26. Prove that, if n is a positive integer,

$$n^n \geqslant 1.3.5 \ldots (2n-1) \geqslant (2n-1)^{\frac{1}{2}n}.$$

27. Prove that $(ax+by)^2 \leqslant (a^2+b^2)(x^2+y^2)$.

28. Find the range of real values of x for which

(i) $x^2-3x+2 > 0$,

(ii) $(x-k)(x^2-3x+2) > 0$,

considering the cases which arise from different real values of k.

29. Prove that the arithmetic mean of two numbers is not less than their geometric mean.

If a, b, c, d are four positive numbers whose sum is 1, find the greatest value of their product.

12

THE GRAPH OF A RATIONAL FUNCTION

1. General features. Many characteristic features of a rational function $f(x)/g(x)$ can be exhibited by means of a sketch of the graph

$$y = \frac{f(x)}{g(x)}.$$

Attention is confined to the cases where $f(x)$, $g(x)$ are constants, linear polynomials or quadratic polynomials, and, for convenience, the exposition is in terms of particular examples. No attempt is made to be exhaustive; the method of procedure is important, but detailed results need not be remembered yet.

What we propose to do is to emphasize by simple examples four general features which serve to settle the nature of the curve. These are:

(i) *The points where* $f(x) = 0$.

When $f(x) = 0$, the value of y is also zero. The curve thus cuts the x-axis at the points arising from the real roots of that equation.

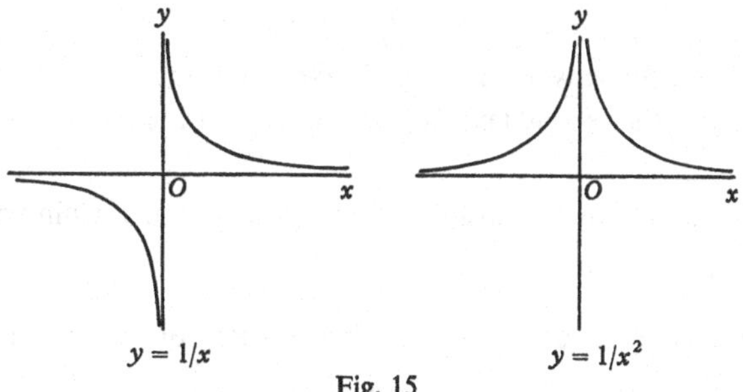

Fig. 15

(ii) *The points where* $g(x) = 0$.

When $g(x) = 0$, the value of y is undefined. A familiar example is the function $1/x$ which is undefined at $x = 0$. The curve

$$y = 1/x$$

has the line

$$x = 0$$

as a 'vertical' asymptote.* A similar example is the curve

$$y = 1/x^2.$$

(See fig. 15.) The implications will be considered later (p. 140).

(iii) *Large values of x.*

For large values of x, positive or negative, the curve may tend towards some readily identifiable line—a 'horizontal' asymptote. Each of the above two curves, $y = 1/x$, $y = 1/x^2$, approaches the line $y = 0$ as x becomes larger and larger.

(iv) *Restrictions on the values of y.*

Corresponding to every value of x, there is defined a single value of y, with the sole exception of those values of x for which $g(x) = 0$. On the other hand, there may be values of y which cannot be attained for any real value of x. Consider, for example, the curve

$$y = \frac{x^2}{x^2+4x+3}.$$

Arranged as a quadratic in x, the equation is

$$(y-1)x^2+4yx+3y = 0.$$

Any given value of y is certainly attained by two values of x, but these values of x are not real unless the condition (p. 20)

$$(4y)^2-4(y-1)(3y) \geqslant 0$$

is satisfied; that is, unless $\quad y^2+3y \geqslant 0,$

or $\quad y(y+3) \geqslant 0.$

Hence y cannot lie between 0 and -3. (If it did, the first term y would be negative and the second $(y+3)$ positive, so that the product would be negative.)

We now examine these features, where relevant, for certain particular curves.

2. Numerator and denominator both without real zeros. Consider the curve

$$y = \frac{x^2+1}{x^2+4}.$$

(i) Since x^2+1 is never zero for real x, the curve does not cut the x-axis; in fact, since $(x^2+1)/(x^2+4)$ is always positive, the curve lies entirely 'above' it.

* For *asymptotes*, see a text-book on calculus or (more restricted in scope) coordinate geometry.

(ii) There are no real values of x for which x^2+4 is zero.

(iii) The behaviour of y for large values of x is ascertained by writing the equation in the form

$$y = \frac{1+(1/x^2)}{1+(4/x^2)}.$$

As x, whether positive or negative, becomes larger and larger, the two functions $1/x^2$ and $4/x^2$ become smaller and smaller. Hence y approaches more and more closely to the value 1. The curve thus has a 'horizontal' asymptote

$$y = 1.$$

(iv) To find whether there are restrictions on the value of y, write the equation in the form

$$x^2(y-1)+(4y-1) = 0.$$

Since x^2 is necessarily positive, this equation cannot be solved unless $y-1$ and $y-\frac{1}{4}$ are opposite the sign. Hence y must lie between 1 and $\frac{1}{4}$. Moreover, there is no solution if $y = 1$ (the *approach* to $y = 1$ was considered previously for large values of x); and the two values of x are equal, each being zero, when $y = \frac{1}{4}$.

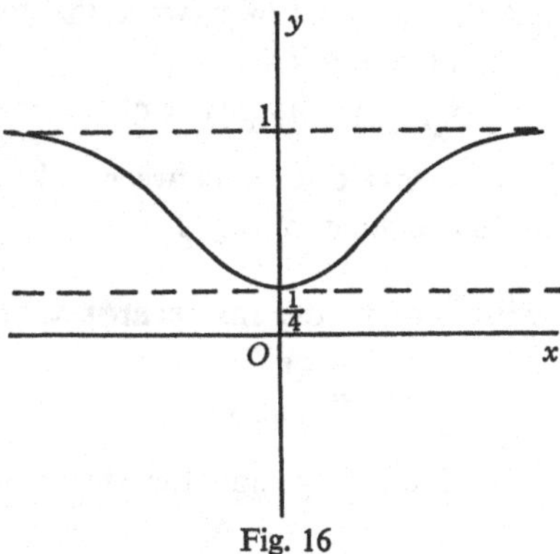

Fig. 16

Thus the curve is restricted to lie between the two lines $y = 1$, $y = \frac{1}{4}$.

These four features are reflected in the diagram (fig. 16), and should be identified carefully. The actual construction of the curve with these 'guides' will be considered later.

3. Numerator with real zeros, denominator without. Consider the curve

$$y = \frac{x^2-1}{x^2+4}.$$

(i) $y = 0$ when $x = 1$ and when $x = -1$.

(ii) There are no real values of x for which x^2+4 is zero.

(iii) For large values of x, write the equation in the form

$$y = \frac{1-(1/x^2)}{1+(4/x^2)}.$$

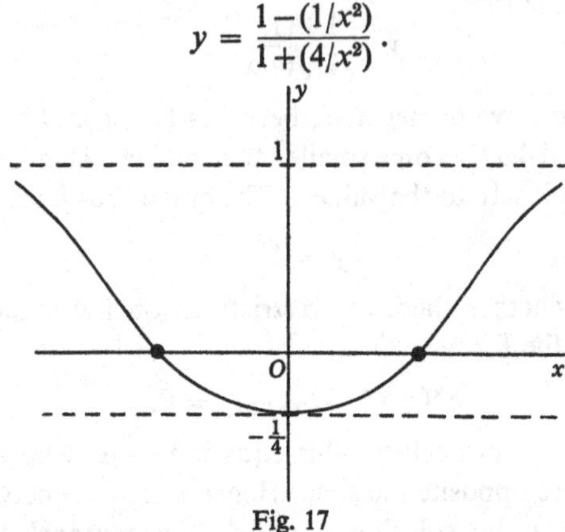

Fig. 17

As in §2, y approaches more and more closely to the value 1 as x becomes larger and larger (positive or negative).

(iv) For restrictions on the value of y, write the given equation, as before, in the form of a quadratic in x:

$$x^2(y-1)+(4y+1) = 0.$$

Since x^2 is positive, y is restricted to lie between $-\frac{1}{4}$ and 1.

The curve is shown in the diagram (fig. 17).

4. Denominator with real zeros, numerator without. Consider the curve

$$y = \frac{x^2+1}{x^2-4}.$$

The difficulties are now increasing, and the steps should be studied carefully.

(i) Since x^2+1 is never zero for real x, the curve does not cut the x-axis.

(ii) The denominator is zero for $x = 2$ and for $x = -2$.

Consider the two cases in turn:

(*a*) $x = 2$.

The function y is not defined when $x = 2$, for the expression

$$\frac{2^2+1}{2^2-4}$$

has no meaning. We therefore seek instead to find what the curve is like *near* to $x = 2$, and, for that purpose, we take a value of x given by

$$x = 2+\frac{1}{k},$$

where k is thought of as some large number (positive or negative) such as 1000 or 1,000,000 or larger.

Since

$$y = \frac{x^2+1}{(x-2)(x+2)},$$

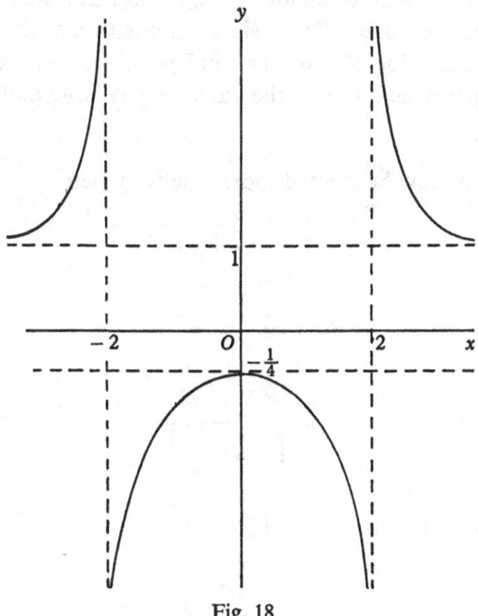

Fig. 18

the value of y corresponding to $x = 2+\frac{1}{k}$ is

$$y = \frac{\left(2+\frac{1}{k}\right)^2+1}{\frac{1}{k}\left(4+\frac{1}{k}\right)}$$

$$= \frac{k\left(5+\frac{4}{k}+\frac{1}{k^2}\right)}{4+\frac{1}{k}}.$$

If the value of x moves closer and closer to 2, then k becomes larger and larger, so that the fraction

$$\frac{5+\dfrac{4}{k}+\dfrac{1}{k^2}}{4+\dfrac{1}{k}}$$

becomes more and more like $\frac{5}{4}$. The value of y can thus be assessed by the approximation

$$\tfrac{5}{4}k,$$

an approximation which becomes steadily closer as k increases, that is, as x approaches the value 2. Thus when k is positive, so that x is slightly larger than 2, the value of y is large and positive; when k is negative, so that x is slightly less than 2, the value of y is large and negative.

(*b*) $x = -2$.

This case may now be treated more briefly. When

$$x = -2+\frac{1}{k},$$

then

$$y = \frac{x^2+1}{(x-2)(x+2)}$$

$$= \frac{5-\dfrac{4}{k}+\dfrac{1}{k^2}}{\left(-4+\dfrac{1}{k}\right)\dfrac{1}{k}}$$

$$= \frac{k\left(5-\dfrac{4}{k}+\dfrac{1}{k^2}\right)}{-4+\dfrac{1}{k}},$$

so that the approximation is

$$y = -\tfrac{5}{4}k.$$

Thus when k is positive, so that x is slightly larger than -2, the value of y is large and negative; when k is negative, so that x is slightly less than -2, the value of y is large and positive.

(iii) For large values of x, we have, as before,

$$y = \frac{1+(1/x^2)}{1-(4/x^2)},$$

so that y approaches the value 1.

(iv) For restrictions on the value of y, write the given equation, as before, in the form

$$x^2(y-1)-(4y+1) = 0.$$

Since x^2 is positive, y is restricted to lie *outside* the range $-\frac{1}{4}, 1$; for values of y between $-\frac{1}{4}$ and 1, the value of $4y+1$ is positive but $y-1$ is negative. Thus *the function*

$$\frac{x^2+1}{x^2-4}$$

takes values greater than $+1$ *and less than* $-\frac{1}{4}$, *but not in between. For any value of y in the allowable range there are two values of x.*

The curve is shown in the diagram (fig. 18), where all these features are exhibited.

5. Numerator and denominator with real zeros, the zeros 'interlacing'.

(We might have considered the curves

$$y = \frac{x^2-1}{x^2-4}, \quad y = \frac{x^2-4}{x^2-1},$$

where the zeros of the numerator lie both inside or both outside those of the denominator. These curves, however, do not provide anything characteristically new. This is the reason for the slight break of pattern from the preceding examples.)

Consider the curve

$$y = \frac{x^2-4x}{x^2-1}.$$

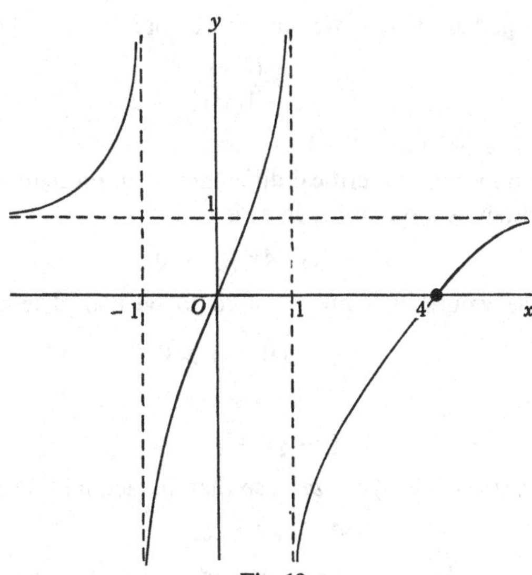

Fig. 19

(i) The numerator is zero when $x = 0$ and when $x = 4$.
(ii) The denominator is zero when $x = 1$ and when $x = -1$.

(*a*) When
$$x = 1+\frac{1}{k},$$
then
$$y = \frac{-3-\dfrac{2}{k}+\dfrac{1}{k^2}}{\dfrac{1}{k}\left(2+\dfrac{1}{k}\right)} = \frac{k\left(-3-\dfrac{2}{k}+\dfrac{1}{k^2}\right)}{2+\dfrac{1}{k}},$$
so that an approximation is
$$y = -\tfrac{3}{2}k.$$
Thus when k is positive (negative), the value of y is large and negative (positive).

(*b*) When
$$x = -1+\frac{1}{k},$$
then
$$y = \frac{5-\dfrac{6}{k}+\dfrac{1}{k^2}}{\left(-2+\dfrac{1}{k}\right)\dfrac{1}{k}} = \frac{k\left(5-\dfrac{6}{k}+\dfrac{1}{k^2}\right)}{-2+\dfrac{1}{k}},$$
so that an approximation is
$$y = -\tfrac{5}{2}k.$$
Thus when k is positive (negative), the value of y is large and negative (positive).

(iii) For large values of x, we have, as before,
$$y = \frac{1-(4/x)}{1-(1/x^2)},$$
so that y approaches the value 1.

(iv) There now comes a critical difference from preceding work. The equation, written as a quadratic in x, is
$$x^2(y-1)+4x-y = 0.$$
Moreover, the values of x are required to be real. The condition is (p. 20)
$$2^2-(y-1)(-y) \geqslant 0,$$
or
$$y^2-y+4 \geqslant 0,$$
or
$$(y-\tfrac{1}{2})^2+\tfrac{15}{4} \geqslant 0.$$
But the least value of $(y-\frac{1}{2})^2$ is zero, so that, in fact, it is always true that
$$y^2-y+4 > 0,$$
and so the *two values of x are real and distinct for each value of y.*

Whatever value of y is selected, say $y = p$, there are two distinct points of the curve which lie on the line $y-p = 0$.

The curve is shown in the diagram (fig. 19), and the reader should verify carefully how all these features are represented.

6. The construction of a curve. Consider the curve

$$y = \frac{x^2-4x+3}{x^2-6x+8}.$$

We take each of the four features (p. 137) in turn, and build up the curve by successive addition of the characters suggested. The letters in the diagrams are inserted for reference at the end.

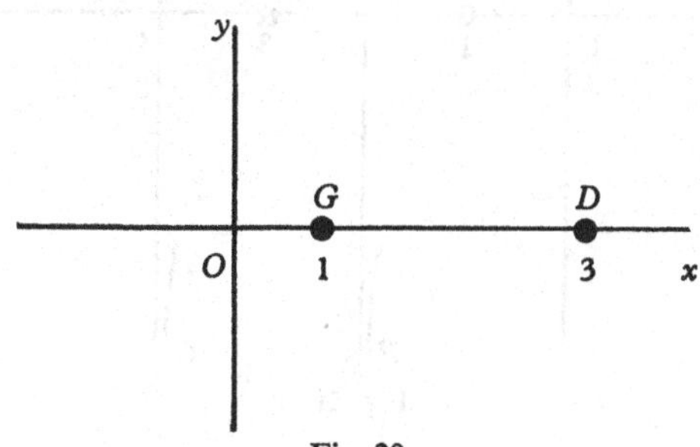

Fig. 20

(i) The numerator is zero when $x = 1$ and when $x = 3$. This fixes two points G, D on the x-axis through which the curve passes (fig. 20).

(ii) The denominator is zero when $x = 2$ and when $x = 4$.

(*a*) When $$x = 2+\frac{1}{k},$$

then $$y = \frac{k\left(-1+\frac{1}{k^2}\right)}{-2+\frac{1}{k}},$$

so that an approximation is $y = \frac{1}{2}k$.

Thus when k is positive (negative), the value of y is large and positive (negative). This settles the two portions of arc at E, F near the line $x = 2$ (fig. 21).

(*b*) When $$x = 4+\frac{1}{k},$$

then

$$y = \frac{k\left(3+\frac{4}{k}+\frac{1}{k^2}\right)}{2+\frac{1}{k}},$$

so that an approximation is $y = \frac{3}{2}k.$

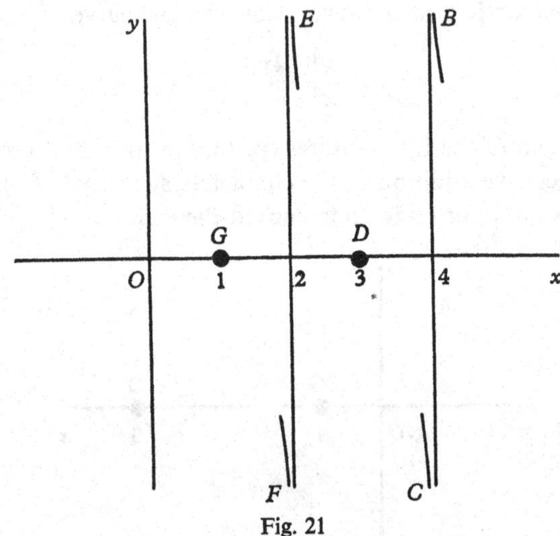

Fig. 21

Thus when k is positive (negative), the value of y is large and positive (negative). This settles the two portions of arc at B, C near the line $x = 4$ (fig. 21).

(iii) For large values of x, the relation is

$$y = \frac{1-\frac{4}{x}+\frac{3}{x^2}}{1-\frac{6}{x}+\frac{8}{x^2}},$$

so that y approaches the value 1. Moreover,

$$y-1 = \frac{\frac{2}{x}-\frac{5}{x^2}}{1-\frac{6}{x}+\frac{8}{x^2}},$$

so that, when x is large and positive, $y-1$ is small and positive; and when x is large and negative, $y-1$ is small and negative. The curve therefore lies 'above' the line $y = 1$ on the right and 'below' it on the left. This is shown at A and H in the diagram (fig. 22).

(iv) To determine whether there are any restrictions on y, write the equation as a quadratic in x in the form

$$x^2(y-1)-2x(3y-2)+(8y-3) = 0.$$

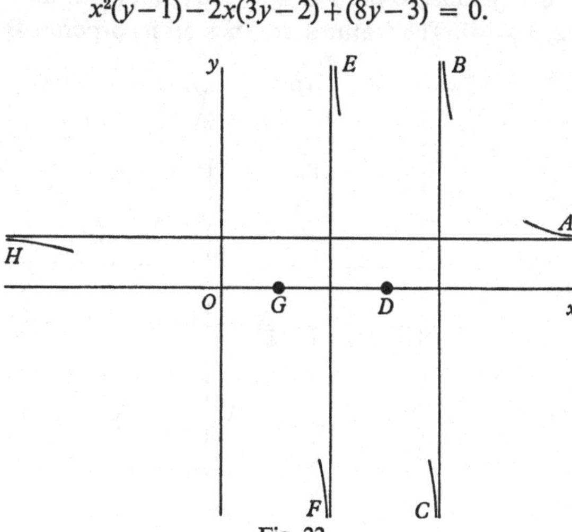

Fig. 22

For each value of y there are two values of x, and these are real provided that

$$(3y-2)^2-(y-1)(8y-3) \geqslant 0,$$

or

$$y^2-y+1 \geqslant 0,$$

or

$$(y-\tfrac{1}{2})^2+\tfrac{3}{4} \geqslant 0.$$

Since $(y-\frac{1}{2})^2$ is always positive, it follows that

$$y^2-y+1 > 0,$$

and so the values of x are real and distinct for all values of y. Hence there are no restrictions on the value of y (save that $y = 1$, exceptionally, is a 'horizontal' asymptote with no genuine intersections with the curve).

The information given by the four features is therefore that indicated in the diagram (fig. 22), and it remains to join the characters up to form the whole curve.

There are now admittedly gaps in the argument, so that the solution is perhaps plausible rather than rigorous, but very limited experience will convince the reader that the resulting sketch (fig. 23) is the only reasonable one.

The 'branch' from A continues up to B; from C, a branch goes through the point D to E; from F a branch goes through the point G to H. The curve is met by every line parallel to the x-axis in two points, and by every line parallel to the y-axis in one point (with the exceptions $y = 1$, $x = 2$, $x = 4$). The features are thus all incorporated.

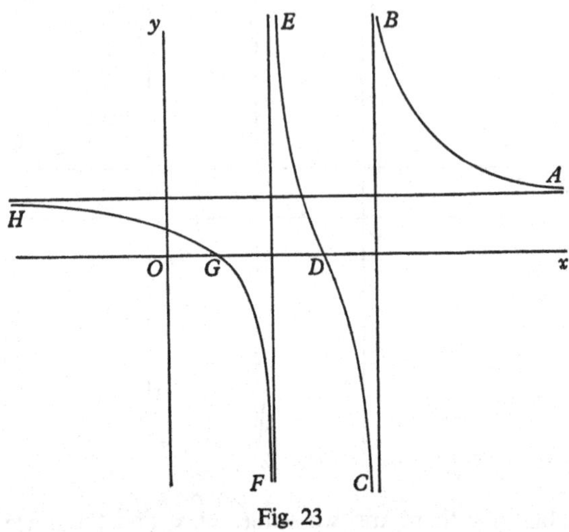

Fig. 23

NOTE. The reader familiar with calculus may construct a more rigorous proof of some of the details. The curve is

$$y = \frac{(x-1)(x-3)}{(x-2)(x-4)},$$

so that, taking logarithms and differentiating,

$$\frac{1}{y}\frac{dy}{dx} = \frac{1}{x-1} + \frac{1}{x-3} - \frac{1}{x-2} - \frac{1}{x-4}.$$

The sign of the gradient may thus be determined for each of the six intervals into which the points $x = 1, 2, 3, 4$ divide the x-axis.

For example, when $x > 4$, y is positive; also

$$\frac{1}{x-1} - \frac{1}{x-2} = \frac{-1}{(x-1)(x-2)} = \text{negative},$$

and
$$\frac{1}{x-3} - \frac{1}{x-4} = \frac{-1}{(x-3)(x-4)} = \text{negative}.$$

Hence dy/dx is negative, so that the curve has negative gradient throughout the range $x > 4$.

Again, when $1 < x < 2$, y is negative; also

$$\frac{-1}{x-2} \text{ is positive, } \frac{-1}{x-4} \text{ is positive,}$$

and
$$\frac{1}{x-1}+\frac{1}{x-3} = \frac{2(x-2)}{(x-1)(x-3)} = \text{positive.}$$

Hence dy/dx is negative, so that the curve has negative gradient throughout the range $1 < x < 2$.

The other intervals may be treated similarly.

Examples 1

Sketch the graphs of the functions:

1. $\dfrac{1}{x-1}$.
2. $\dfrac{1}{(x-2)^2}$.
3. $\dfrac{x}{x-1}$.
4. $\dfrac{x-1}{x-2}$.
5. $\dfrac{x-1}{x}$.
6. $\dfrac{x}{x^2+1}$.
7. $\dfrac{x-1}{x^2+1}$.
8. $\dfrac{x^2+1}{x^2+2}$.
9. $\dfrac{x^2-1}{x^2}$.
10. $\dfrac{x^2}{(x-1)(x-2)}$.
11. $\dfrac{x^2-x}{x^2+4}$.
12. $\dfrac{x^2+1}{x^2-x}$.

Revision Examples VII

1. Prove that as x varies the function

$$\frac{(x-1)(x-5)}{2x-1}$$

assumes all real values except those lying between -4 and -1, and illustrate this result by drawing a rough graph of the function.

2. Determine the ranges of the values of the expression

$$\frac{(x+c)^2}{x+1}$$

when x may have all real values, distinguishing the cases $c > 1$, $c = 1$, $c < 1$.

3. Show that, if $$y = \frac{x-1}{(x-2)(x-3)},$$

then x is not real when y has any value between $-3-2\sqrt{2}$ and $-3+2\sqrt{2}$.

4. Show that, if
$$y = \frac{x-p}{(x-q)(x-r)},$$
then x is real for every real value of y provided that
$$(p-q)(p-r) \leqslant 0.$$

5. Prove that, if x is real, the value of
$$\frac{x^2-22x-47}{x^2+10x+17}$$
cannot lie between -3 and -7.

6. Show that, if x is real, the expression
$$\frac{240x}{x^2+2400}$$
lies between $-\sqrt{6}$ and $+\sqrt{6}$.

Draw a graph of the function, and use it to deduce the number of real roots of the equation
$$(x^2+2400)(x^2-2)+240x = 0.$$

7. Sketch the graph of the function
$$y = \frac{x^2-4x+3}{x^2-2x}.$$

8. Sketch the curve $\quad y = \dfrac{x^2}{x^2+3x+2}.$

By means of the line $\quad y+8 = m(x+1),$

or otherwise, find the number of real roots of the equation
$$m(x+1)^2(x+2) = (3x+4)^2,$$
when m is a real constant which is (i) positive, (ii) negative.

9. Find the limitations on the value of a in order that the rational function
$$\frac{x^4+4x-5}{x^2+2x+a}$$
may take every real value for real values of x.

Determine the restriction on the values of the function, for all real values of x, when a does not satisfy these limitations.

10. Draw rough graphs illustrating the behaviour of the function
$$\frac{(x-a)(x-b)}{x}$$
for different values of a, b.

11. Prove that for real values of x the rational function

$$\frac{5x^2-18x-35}{8(x^2-1)}$$

takes all real values except those between 1 and 4.

Draw a rough graph of the function.

12. If
$$y=\frac{x-1}{(x+1)^2},$$
show that y can never be greater than $\frac{1}{8}$, and sketch the graph.

13. If A, B, α, β are real constants and x a real variable, show that

$$\frac{A}{x-\alpha}+\frac{B}{x-\beta}$$

can take all real values if A, B have the same sign; but that there is a range of values, of extent $4\sqrt{(-AB)}/|\alpha-\beta|$, which the expression cannot take if A, B have opposite signs and $\alpha \neq \beta$.

14. Draw rough sketches to illustrate the behaviour of the curve

$$y=\frac{x^2-1}{x-a}$$

for different values of a.

15. Prove that, for real values of x, the function

$$\frac{3x^2+18x+15}{x^2+10x+7}$$

takes all real values except those in a certain range (to be found).

Sketch the graph of the function.

16. Prove that the function

$$\frac{x^2-2x+21}{6x-14}$$

can take, for real x, all values except those lying between $-\frac{10}{9}$ and 2.

17. Prove that the function

$$\frac{kx+1}{x^2+4x+3}$$

assumes all values as x ranges from $-\infty$ to $+\infty$, provided that $\frac{1}{3}<k<1$.

18. If
$$y = \frac{8x-2}{2x^2+1},$$
prove that, when x is real, y can have any value between -4 and 2 inclusive, but no other.

Sketch roughly the graph of the function.

19. Show that the values of the function
$$\frac{x^2+x+1}{x^2+1},$$
for real values of x, lie between $\frac{1}{2}$ and $\frac{3}{2}$.

20. Prove that, if the two equations
$$ax^2+2bx+c = 0,$$
$$a'x^2+2b'x+c' = 0$$
have a single common root, then
$$4(b^2-ac)(b'^2-a'c')-(ac'+a'c-2bb')^2 = 0.$$

Show that the condition
$$4(b^2-ac)(b'^2-a'c')-(ac'+a'c-2bb')^2 \geqslant 0$$
is necessary for the fraction
$$\frac{ax^2+2bx+c}{a'x^2+2b'x+c'}$$
to assume all real values for real values of x; and that, if this condition is not fulfilled, the range of inadmissible values of the fraction are either entirely between or entirely outside the roots of the equation
$$(b'^2-a'c')x^2+(ac'+a'c-2bb')x+b^2-ac = 0.$$

21. Prove that, if all the numbers involved are real, the function
$$f(x) \equiv \frac{ax^2+2bx+c}{x^2+k}$$
is capable of all real values if
$$a^2k^2+2k(2b^2-ac)+c^2 < 0.$$

Prove that this inequality implies the two
$$b^2-ac > 0, \quad k < 0,$$
and investigate the conditions for the existence of two limiting values between which (i) $f(x)$ cannot lie, (ii) $f(x)$ must lie.

13

PERMUTATIONS AND COMBINATIONS

We do not propose to go into much detail about problems on *choice and chance*. The brief account which follows is devoted chiefly to those aspects which are required for work on the binomial theorem.

1. Permutations; objects distinguishable. The essential permutation problem is:

Given a certain number, say n, of distinct objects, in how many different ways can an arrangement be made involving p of them?

Each such arrangement is called a *permutation.*

The procedure may be illustrated by particular examples.

Illustration 1. *To find the number of* 3-*digit numbers that can be formed from the digits* 1, 2, 3, 4, 5 (*without repetitions*).

The first digit can be selected in 5 ways; then the second in 4 ways; then the third in 3 ways. The total number of permutations is therefore

$$5\times4\times3,$$

or 60.

The numbers chosen can be exhibited by the scheme:

```
123, 124, 125    213, 214, 215    312, 314, 315
132, 134, 135    231, 234, 235    321, 324, 325
142, 143, 145    241, 243, 245    341, 342, 345
152, 153, 154    251, 253, 254    351, 352, 354

         412, 413, 415    512, 513, 514
         421, 423, 425    521, 523, 524
         431, 432, 435    531, 532, 534
         451, 452, 453    541, 542, 543
```

There are 5 blocks corresponding to the five choices of first digit; within each block are 4 rows corresponding to the four choices of second digit; within each row are 3 groups, corresponding to the three choices of third digit.

Illustration 2. *To find the number of ways in which an arrangement can be made involving* 4 *of the letters a, b, c, d, e, f, g.*

An arrangement can be built up by successive operations:

(i) The first letter may be any one of the 7 letters, and so may be chosen in 7 ways;

(ii) Once the first letter is chosen, the second may be chosen in 6 ways; so that the first 2 letters may be chosen altogether in 7×6 ways.

(iii) Once the first 2 letters are chosen, the third may be chosen in 5 ways; so that the first 3 letters may be chosen altogether in $7\times 6\times 5$ ways.

(iv) Once the first 3 letters have been chosen, the fourth may be chosen in 4 ways; so that the first 4 letters may be chosen altogether in $7\times 6\times 5\times 4$ ways.

Hence the number of permutations is

$$7\times 6\times 5\times 4,$$

or 840.

The general formula may now be established:

The number of permutations of n distinct objects taken p at a time is

$$n(n-1)(n-2)\ldots(n-p+1).$$

NOTATION. The number of such permutations is denoted by the symbol ${}_nP_p$, so the formula to be proved is

$${}_nP_p = n(n-1)(n-2)\ldots(n-p+1).$$

[For example, the two Illustrations establish the formulae

$${}_5P_3 = 5.4.3 = 60,$$

$${}_7P_4 = 7.6.5.4 = 840.]$$

The first object can be selected in n ways; then the second in $n-1$ ways; then the third in $n-2$ ways; and so on, for p steps. The number of permutations is thus, taking one from each selection,

$$n\times(n-1)\times(n-2)\times\ldots$$

to p factors. Now the pth factor is $n-p+1$, so that

$${}_nP_p = n(n-1)(n-2)\ldots(n-p+1).$$

COROLLARY. The number of ways in which n objects can be permuted among themselves (all objects being present in each permutation) is

$$n(n-1)(n-2)\ldots 3.2.1.$$

NOTATION. The product

$$n(n-1)(n-2)\ldots 3.2.1$$

is denoted by the symbol $n!$, called '*n factorial*'. With this notation,

$${}_nP_p = \frac{n\ldots(n-p+1)(n-p)\ldots 1}{(n-p)\ldots 1} = \frac{n!}{(n-p)!}.$$

There are two modifications which may be mentioned, though they are not directly relevant here.

2. Cyclic permutations. It was implicit in §1 that the objects, when selected, were arranged, as it were, in a straight line. A slight change is required when they are arranged round a circle without characteristics.

Illustration 3. *To find the number of ways in which the six points A, B, C, D, E, F can be placed on the circumference of a circle so that consecutive members are equidistant.*

The point A, say, may be placed at random on the circle. The position of A then characterizes the rest of the circle (any point of which may be defined by its distance from A in the counterclockwise sense). The number of permutations is therefore the number of ways in which the remaining 5 points can be arranged, namely

$$5 \times 4 \times 3 \times 2 \times 1,$$

or 120.

3. Permutations; objects not all distinguishable. The number of permutations is reduced when the objects cannot all be distinguished from each other.

Illustration 4. *To find the number of* 6-*digit numbers that can be formed from the digits* 1, 2, 3, 4, 4, 4.

(i) If we regard the three 4's as temporarily distinct, the number of permutations is

$$6 \times 5 \times 4 \times 3 \times 2 \times 1.$$

(ii) Given any one permutation, the three 4's in it may be re-arranged among themselves, in

$$3 \times 2 \times 1$$

ways, without altering the 6-digit number.

(iii) Hence the number of distinct 6-digit numbers is

$$\frac{6 \times 5 \times 4 \times 3 \times 2 \times 1}{3 \times 2 \times 1} \equiv \frac{6!}{3!},$$

or 120.

Illustration 5. *To find the number of* 10-*digit numbers that can be formed from the digits* 1, 1, 1, 2, 2, 2, 2, 3, 4, 5.

(i) If we regard the 1's and 2's as temporarily distinct, the number of permutations is

$$10!.$$

(ii) Given any one permutation, the three 1's may be re-arranged among themselves in 3! ways, and the four 2's may be re-arranged among themselves in 4! ways, without altering the 10-digit number.

(iii) Hence the number of distinct 10-digit numbers is

$$\frac{10!}{3!\,4!},$$

or 25,200.

***Illustration* 6.** *To find the number of 4-letter 'words' that can be made from a choice of the letters a, a, a, b, c, d, e, f.*

The 'word' may contain 0, 1, 2, or 3 of the letters *a*. Using the principles of the preceding Illustrations, we have the following results:

(i) *No letter a.* The 'word' contains 4 of *b, c, d, e, f*, and the number is

$$5.4.3.2,$$

or 120.

(ii) *One letter a.* The 'word' contains *a*, which can be placed in 4 ways; and then 3 of *b, c, d, e, f* which can be chosen in 5.4.3 ways. The total is

$$4\times 5.4.3,$$

or 240.

(iii) *Two letters a.* The 'word' contains *a, a*, which can be placed in (4.3)/(2.1), or 6 ways; and then 2 of *b, c, d, e, f* which can be chosen in 5.4 ways. The total is

$$\frac{4.3}{2.1}\times 5.4,$$

or 120.

(iv) *Three letters a.* The 'word' contains *a, a, a*, which can be placed in (4.3.2)/(3.2.1), or 4 ways; and then 1 of *b, c, d, e, f*, which can be chosen in 5 ways. The total is

$$\frac{4.3.2}{3.2.1}\times 5,$$

or 20.

Hence the total number of 'words' is

$$120+240+120+20,$$

or 500.

4. Combinations. The essential combination problem is:

Given a certain number, say n, of objects, in how many different ways can a group be chosen containing p of them? The order of the objects within the group is regarded as irrelevant.

Each such group is called a *combination.*

The procedure may be illustrated by particular examples:

***Illustration* 7.** *To find the number of different numbers that can be formed by multiplying together any four of the numbers* 2, 3, 5, 7, 11, 13, 17, 19, 23.

We must first select the four numbers. They can be written down in

$$9.8.7:6$$

ways. But in forming the product, the order in which the four numbers are arranged among themselves is irrelevant (and they are mutually prime, so all products are different), so that any one product comes from 4! of the above ways. Hence the number of *distinct* products is

$$\frac{9.8.7.6}{4.3.2.1},$$

or 126.

***Illustration* 8.** *To find the number of different ways in which a group of three may be selected from the letters a, b, c, d, e, f, g.*

We must first select the three letters. They can be written down in

$$7.6.5$$

ways. But in the forming of a group the order is irrelevant, so that any one group comes from 3! of the above ways. Hence the number of *distinct* groups is

$$\frac{7.6.5}{3.2.1},$$

or 35.

The general formula may now be established:

The number of combinations of n objects taken p at a time is

$$\frac{n!}{p!(n-p)!}.$$

NOTATION. The number of such combinations is denoted by the symbol ${}_nC_p$, so that the formula to be proved is

$${}_nC_p = \frac{n!}{p!(n-p)!}.$$

[For example, the two Illustrations establish the formulae

$${}_9C_4 = \frac{9!}{4!\,5!} = 126,$$

$${}_7C_3 = \frac{7!}{3!\,4!} = 35].$$

Select first the p objects. They can be put into *ordered* sets of p in ${}_nP_p$ ways. But in forming a group, the order is in fact irrelevant, so that any one group comes from $p!$ of the above ways. Hence the number of distinct groups is

$$\frac{{}_nP_p}{p!}.$$

But ${}_nP_p = n!/(n-p)!$, so that

$${}_nC_p = \frac{n!}{p!(n-p)!}.$$

Note that

$${}_nC_{n-p} = \frac{n!}{(n-p)!\,\{n-(n-p)\}!} = \frac{n!}{(n-p)!\,p!}.$$

$$= {}_nC_p.$$

In fact, the number of combinations formed by *omitting* p objects (thus selecting $n-p$) is the same as the number formed by *including* p of them.

5. Combinations; several groups. The ideas of §§3, 4 are closely related, as an alternative proof of the formula for ${}_nC_p$ shows:

Given n objects, the selection of a group of p of them is effected by dividing the set into two classes, one of p objects and the other of q objects, where

$$p+q = n,$$

and then accepting the first class and consequently rejecting the second. The total number of ways in which all n objects may be arranged is $n!$. But in forming a group of p objects, their $p!$ arrangements among themselves are irrelevant; and in forming a group of q objects, their $q!$ arrangements among themselves are irrelevant. Hence the number of distinct ways of forming the two classes is

$$\frac{n!}{p!\,q!},$$

and this, since $q = n-p$, is the formula for ${}_nC_p$ already obtained.

This reasoning can be extended. Suppose that there is a set of n objects to be divided into a number of classes containing $p, q, r, \ldots$ objects per class, where

$$p+q+r+\ldots = n.$$

Once any of the possible $n!$ arrangements has been made, the $p!$ re-arrangements of the p objects selected for the first class, the $q!$ re-arrangements of the q objects selected for the second class, ..., are irrelevant. Hence *the number of distinct ways of forming the classes is*

$$\frac{n!}{p!\,q!\,r!\ldots},$$

where $p+q+r+\ldots = n$.

***Illustration* 9.** *A number of objects are labelled* 1, 2, 3, 4, 5, 6, 7, 8, 9. *In how many ways is it possible to place* 2 *in one bag,* 3 *in another bag, and* 4 *in another?*

The objects may be imagined as laid out in a line ready for selecting. This can be done in 9! ways. The first 2 are selected, then the next 3, then the next 4. The 2! re-arrangements of the first two objects among themselves will not disturb the selection; neither will the 3! of the next 3, or the 4! of the next 4. Hence the number of distinct selections is

$$\frac{9!}{2!\,3!\,4!},$$

or 1260.

6. Probability. Any detailed examination of probability would carry us much too far, but one or two elementary examples may be given.

Suppose that a given event can happen in N ways, and that one particular selection from those events can happen in n of those N ways. Then the *probability* or *chance* of that particular selection is defined to be the ratio n/N.

(i) *A bag contains three red balls, four yellow balls, and five black balls. To find the probability that, if a ball is drawn at random from the bag, its colour will be black.*

The total number of ways of drawing a ball is $3+4+5 = 12$. The number of ways of drawing a black ball is 5. Hence the probability is

$$\frac{5}{12}.$$

(ii) *A two-digit number is formed from the integers* 1, 2, 3, ..., 9. *To find the probability that the sum of the digits is less than* 6.

Each of the nine integers chosen first can be combined with each of the nine integers chosen second, to a total of 81 possibilities.

A sum less than six is obtained from

1 1
1 2; 2 1
1 3; 2 2; 3 1
1 4; 2 3; 3 2; 4 1.

This gives 10 possibilities, so the probability is

$$\frac{10}{81}.$$

(iii) *Eight identical notices are put at random in six different pigeon-holes. To find the chance that* 3 *notices are in the first pigeon-hole.*

The first notice may (in the absence of restrictions) be placed in 6 ways; the second in 6 independent ways, giving 6^2 so far; the third in 6 further independent ways, giving 6^3 so far; and so on. The total number of possible placings is

$$6^8.$$

Now the three notices to be put in the first pigeon-hole may be selected in ${}_8C_3$ ways. When this has been done, the other 5 may be put into the remaining 5 holes in 5^5 ways. Hence the number of possible ways is

$${}_8C_3 . 5^5 = 56 \times 5^5.$$

Hence the chance (probability) is

$$\frac{56 \times 5^5}{6^8}.$$

Examples 1

1. How many numbers can be formed using all of the digits

(i) 1, 2, 3, 4, 5, (ii) 1, 1, 1, 2, 3, 4, (iii) 1, 1, 1, 1, 2, 2, 2, 3?

2. How many distinct sums can be formed by adding (i) 3, (ii) 6 of the numbers 1, 2, 4, 8, 16, 32, 64, 128, 256?

3. How many committees of 9 members can be chosen from 8 men and 6 women (i) without restriction, (ii) with exactly 1 woman, (iii) with not less than 3 women?

4. A number less than 50 is selected at random. What is the probability that it is (i) a multiple of 3, (ii) a multiple of 4, (iii) less than 12, (iv) greater than 38?

5. A card is drawn at random from a normal pack of 52. What is the probability that it is (i) a king, (ii) a heart, (iii) the king of hearts?

6. In how many ways can 10 representatives sit at a round table (without characteristic place)? What is the number if A and B must always sit (i) together, (ii) apart?

7. How many triangles can be selected having their vertices at three points chosen from (i) A, B, C, D, (ii) A, B, C, D, E, (iii) A, B, C, D, E, F, G, H?

8. A committee of 4 is chosen from 3 men and 5 women. What is the probability that there are (i) 2 men, (ii) 3 men, (iii) 0 men?

Revision Examples VIII

1. A number N has its five digits selected from the digits 1, 2, 3, 4, 5 without repetitions. How many such numbers N are there?
In how many of these numbers is 3 the first digit?

2. In how many ways can a committee of 4 be chosen from 6 men and 3 women (i) without any restriction, (ii) if there must always be just one woman on the committee?

3. In how many ways can four cards be selected from a pack of 52 cards, if at least one of them is to be an ace?

4. A pair of dice is thrown. What is the probability that a 2 and a 3 turn up? What is the probability that the sum of the two numbers turning up is 7?

5. How many numbers of 5 digits do not contain a nought? How many of these do not include the same digit more than three times?

6. A bag contains the ten numbers 0, 1, 2, ..., 9. Three numbers are drawn simultaneously. Show that the probability of the three numbers adding up to 13 is 1/12.

7. Ten coloured beads are to be arranged in a circle. In how many ways can this be done: (i) when the beads are all of different colours, (ii) when three are of the same colour and the rest are different?

8. How many factors has the number

$$2^9.3^5.5^4$$

including 1 and the number itself?

9. If n different objects are placed round a ring, prove that the number of ways of selecting 3 of them, so that no selection contains two adjacent objects, is

$$\tfrac{1}{6}n(n-4)(n-5).$$

10. Four articles are distributed to four persons, with no restrictions as to how many any person may receive. Show that the probability that 2 will be given to one person and 1 to each of two others is 9/16.

11. A pack of 52 cards is dealt in the usual way to 4 players. Find an expression for the number of ways in which exactly 6 diamonds may be dealt to a particular player, and show that, in 4 out of every 35,853 of the deals, his partner would have the remaining 7 diamonds.

12. Prove that the number of ways n different letters can be arranged in a row is $n!$

Prove also that the number of 'sentences' of r 'words' and n letters which can be formed from them is

$$\frac{n!\,(n-1)!}{(n-r)!\,(r-1)!}.$$

13. A number, n, of points in a plane are joined in all possible ways by straight lines, produced in both directions; no two of the lines are coincident or parallel, and no three pass through the same point. Prove that the number of points of intersection, exclusive of the n original points, is

$$\tfrac{1}{8}n(n-1)(n-2)(n-3).$$

14. A number p of objects are put at random in n different cells. Prove that the chance that k objects are in any particular cell is

$$\frac{p!}{k!\,(p-k)!}\,\frac{(n-1)^{p-k}}{n^p},$$

where $k < p$.

15. Prove that the total number of ways in which a distinct set of three non-zero positive integers can be chosen so that the sum is a given odd integer p is the integer nearest to $\frac{1}{12}p^2$.

16. A match between two players A, B is won by whoever first wins n games. A's chances of winning, drawing or losing any particular game are p, q, r respectively. Prove that his chance of winning the match when n is 2 is

$$\frac{p^2(p+3r)}{(p+r)^3},$$

and of winning when n is 3 is

$$\frac{p^3(p^2+5pr+10r^2)}{(p+r)^5}.$$

17. Find the number of ways of distributing 12 similar coins among seven persons so that at least two persons receive none.

18. Seven slips of paper, three red and four blue, are placed in a bag. Show that, if three slips are drawn at random from the bag, the chances are 6 to 1 against all three having the same colour.

19. If $p > q-2$, find the number of ways in which p positive and q negative signs can be placed in a row so that no two negative signs are together.

20. If p is a positive integer, show that the number of distinct ways in which four positive (non-zero) integers can be chosen to have a sum $12p+1$, and such that two and only two of them are equal, is $p(18p-7)$.

21. Four pennies and four half-crowns are placed in a row on a table. Find the chance that (i) no two pennies are adjacent; (ii) no two heads are adjacent (and showing).

22. A number n of identical cards are to be placed in p pigeon-holes. Show that the number of ways in which the cards can be disposed is

$$\frac{(n+p-1)!}{(p-1)!\,n!}.$$

23. Show that a sum of n pence can be made up of pennies, half-pennies and farthings in $(n+1)^2$ different ways.

24. Points, n in number, are taken in space, so that not more than three points lie in any plane and not more than two in any straight line. Show that these points define $\frac{1}{2}n(n-1)$ straight lines and $\frac{1}{2}\{(n-1)!\}$ polygons (skew) of n sides.

25. A pack of cards is dealt in the usual way to four players. One player has 5 diamonds. Prove that the chance that his partner has the remaining 8 diamonds is

$$1/(4.17.19.37).$$

26. Given a_1 things of one kind, a_2 of another, a_3 of another, and a_4 of a fourth, show that the number of different groups which can be formed from one or more of these things is

$$(a_1+1)(a_2+1)(a_3+1)(a_4+1)-1.$$

Show also that, if $a_1 < n < a_2 < a_3 < a_4$, the number of groups having n numbers is

$$\tfrac{1}{6}\{(n+1)(n+2)(n+3)-(n-a_1)(n-a_1+1)(n-a_1+2)\}.$$

27. Given a series of 5 letters, show that the number of ways in which the order may be deranged so that neither of two *assigned* letters occupies its original place is

$$5!-2(4!)+3!.$$

Prove that the corresponding result for a series of n letters is

$$n!-2(n-1)!+(n-2)!$$

28. It is required to find sets of three positive integers (none zero) whose sum is an even integer $2n$. Prove that, when $n = 2, 3, 4, 5$, the number of distinct sets is 1, 3, 5, 8.

Prove also that, for a general value of n, the number of distinct sets is the integral part of $\frac{1}{3}n^2$.

29. Ten pennies are to be distributed among five people. Prove that there are 875 ways of making the distribution so that at least one person gets nothing.

30. Show that the number of different ways of distributing $p+n$ similar coins among p people so that each person gets at least one coin is

$$\frac{(p+n-1)!}{n!\,(p-1)!}.$$

31. Show that the number of combinations $n-2$ at a time of n things of which 3 are alike and the rest different is

$$\tfrac{1}{2}(n^2-5n+8).$$

32. Show that the number of combinations $n-3$ at a time of n things of which 4 are alike and the rest different is

$$\tfrac{1}{3}(n-3)(n^2-9n+26).$$

33. Ten articles are to be placed in a row, three of them, A, B, C coming together. Prove that this can be done in 241,920 ways.

In how many ways can 9 articles be arranged in a row so that two of them, A and B, do not come together?

34. Four cards are drawn at random from a pack of 52 cards. Show that the probability that they will all be spades is 11/4165.

35. In how many ways can m shillings and n half-crowns be arranged in a line?

In how many of these arrangements are both the extreme coins half-crowns?

Deduce the probability that the extreme coins are both half-crowns.

36. A hand of 13 cards is dealt from an ordinary pack of 52 playing cards. Show that the probability of the hand containing all four aces is 11/4165.

37. A till contains six half-crowns, four florins, two sixpences and twenty-four pennies. How many different selections of coins can be taken from the till?

38. Each of 10 points in one of a given pair of parallel straight lines is joined to each of 21 points in the other by straight lines terminated at the points. If the points are such that no three of the constructed lines are concurrent, prove that the lines intersect in 9450 points, excluding the 31 given points.

39. Prove that the number of ways of dividing $x+y+z$ different objects into three groups containing x, y, z objects respectively is

$$\frac{(x+y+z)!}{x!\,y!\,z!}.$$

Find how many different hands can be held by four people playing bridge (13 cards each), stating the answer in factorial form.

14

THE BINOMIAL THEOREM

1. The product $(x+a)(x+b)\ldots(x+k)$. We can obtain, by repeated multiplication, the formulae

$$\begin{aligned}
&(x+a)(x+b)\\
&\qquad \equiv x^2+(a+b)x+ab,\\
&(x+a)(x+b)(x+c)\\
&\qquad \equiv x^3+(a+b+c)x^2+(bc+ca+ab)x+abc,\\
&(x+a)(x+b)(x+c)(x+d)\\
&\qquad \equiv x^4+(a+b+c+d)x^3+(bc+ca+ab+ad+bd+cd)x^2\\
&\qquad\quad +(bcd+cad+abd+abc)x+abcd,
\end{aligned}$$

wherein the expansions are grouped to form polynomials in x of degree equal to the number of factors.

Consider the product $(x+a)(x+b)(x+c)$ and its expansion.

(i) The terms in x^2 are found by the rule: take the term x from two of the brackets, and *therefore* the term without x from the third; do this in all possible ways. This gives $x^2a+x^2b+x^2c$.

(ii) The terms in x are found by the rule: take the term x from one of the brackets, and *therefore* the term without x from the two others; do this in all possible ways. This gives $xbc+xca+xab$.

Similar methods may be used for the more elaborate products. For example, the coefficient of x^2 in $(x+a)(x+b)(x+c)(x+d)$ is found by the rule: take the term x from two of the brackets, and *therefore* the term without x from the two others; do this in all possible ways. This gives

$$x^2bc+x^2ca+x^2ab+x^2ad+x^2bd+x^2cd.$$

It may help to clarify ideas if the selection of these six terms in succession is indicated by the scheme:

$$\begin{gathered}
(x+\cdot)(\cdot+b)(\cdot+c)(x+\cdot),\\
(\cdot+a)(x+\cdot)(\cdot+c)(x+\cdot),\\
(\cdot+a)(\cdot+b)(x+\cdot)(x+\cdot),\\
(\cdot+a)(x+\cdot)(x+\cdot)(\cdot+d),\\
(x+\cdot)(\cdot+b)(x+\cdot)(\cdot+d),\\
(x+\cdot)(x+\cdot)(\cdot+c)(\cdot+d).
\end{gathered}$$

Proceeding in this way, the general rule for the formation of the coefficients is obtained:

When the product

$$(x+a)(x+b)(x+c)\ldots(x+k)$$

of n factors is expanded as a polynomial in x, the coefficient of x^p consists of the sum of products, each of $n-p$ letters, chosen in all possible ways (without repetitions of letters in any one product) *from a, b, c, ..., k.*

The number of terms in the coefficient of x^p is the number of ways in which $n-p$ letters can be chosen from n, order being irrelevant. This number is

$${}_nC_{n-p},$$

or

$$\frac{n!}{(n-p)!\,p!}.$$

This number is also (p. 157) ${}_nC_p$.

Examples 1

Expand the following products:

1. $(x+1)(x+2)(x+3)$.
2. $(x+1)(x-2)(x+3)$.
3. $(x-1)(2x-1)(3x-1)$.
4. $(x+1)(x-1)(3x+4)$.
5. $(x+1)(x+2)(x+3)(x+4)$.
6. $(x+1)(x-2)(x-3)(x+4)$.
7. $(2x-1)(2x-2)(2x-3)(2x-4)$.
8. $(3x-1)(2x-1)(x-1)(x+1)$.
9. $(x+a)(x+2b)(x+3c)$.
10. $(ax+b)(ax-b)(bx+a)$.

2. The binomial theorem. To prove that, *when n is a positive integer,*

$$(x+a)^n = x^n + {}_nC_1\,x^{n-1}a + {}_nC_2\,x^{n-2}a^2 + \ldots + {}_nC_p\,x^{n-p}a^p + \ldots + {}_nC_n\,a^n.$$

The multiplication of $(x+a)^n$ to form a polynomial in x is the same as that of

$$(x+a)(x+b)(x+c)\ldots(x+k)$$

to n factors when $a, b, c, \ldots, k$ are all put equal to a. By § 1, the coefficient of x^p in the latter consists of ${}_nC_{n-p}$ terms, and each of them (when $a = b = c = \ldots = k$), is a^{n-p}, so that the coefficient is, in all,

$${}_nC_{n-p}\,a^{n-p}.$$

Thus the coefficients of $x^{n-1}, x^{n-2}, x^{n-3}, \ldots$ are

$${}_nC_1a,\ {}_nC_2a^2,\ {}_nC_3a^3, \ldots.$$

Hence, adding all such terms,

$$\begin{aligned}(x+a)^n &= x^n + {}_nC_1x^{n-1}a + {}_nC_2x^{n-2}a^2 + \ldots + {}_nC_px^{n-p}a^p + \ldots + {}_nC_na^n\\ &= x^n + nx^{n-1}a + \frac{n(n-1)}{2!}x^{n-2}a^2 + \frac{n(n-1)(n-2)}{3!}x^{n-3}a^3 + \ldots.\end{aligned}$$

Note that the arithmetical parts of the coefficients are equal at equal distances from the ends.

For example,

$$(x+a)^2 \equiv x^2+2ax+a^2,$$
$$(x+a)^3 \equiv x^3+3ax^2+3a^2x+a^3,$$
$$(x+a)^4 \equiv x^4+4ax^3+6a^2x^2+4a^3x+a^4,$$
$$(x+a)^5 \equiv x^5+5ax^4+10a^2x^3+10a^3x^2+5a^4x+a^5.$$
$$(x+a)^6 \equiv x^6+6ax^5+15a^2x^4+20a^3x^3+15a^4x^2+6a^5x+a^6.$$

Illustration 1. Expressions of greater complexity may be expanded by obvious extensions of the theorem. The examples which follow are typical.

(i) $(x-a)^4 \equiv x^4-4ax^3+6a^2x^2-4a^3x+a^4,$

the signs alternating from term to term.

(ii) $$(x+5)^3 \equiv x^3+3.5x^2+3.5^2x+5^3$$
$$\equiv x^3+15x^2+75x+125.$$

(iii) $$(2x-3)^7 \equiv (2x)^7-7(2x)^6(3)+21(2x)^5(3)^2-35(2x)^4(3)^3$$
$$+35(2x)^3(3)^4-21(2x)^2(3)^5+7(2x)(3)^6-(3)^7$$
$$\equiv 128x^7-1344x^6+6048x^5-15{,}120x^4$$
$$+22{,}680x^3-20{,}412x^2+10{,}206x-2187.$$

(iv) $$\left(x+\frac{1}{x}\right)^4 \equiv x^4+4x^3\left(\frac{1}{x}\right)+6x^2\left(\frac{1}{x}\right)^2+4x\left(\frac{1}{x}\right)^3+\left(\frac{1}{x}\right)^4$$
$$\equiv x^4+4x^2+6+\frac{4}{x^2}+\frac{1}{x^4}.$$

(v) $$\left(x^2-\frac{2}{x}\right)^5 \equiv x^{10}-5x^8\left(\frac{2}{x}\right)+10x^6\left(\frac{2}{x}\right)^2-10x^4\left(\frac{2}{x}\right)^3+5x^2\left(\frac{2}{x}\right)^4-\left(\frac{2}{x}\right)^5$$
$$\equiv x^{10}-10x^7+40x^4-80x+\frac{80}{x^2}-\frac{32}{x^5}.$$

(vi) $$(1+2x+3x^2)^3 \equiv 1+3(2x+3x^2)+3(2x+3x^2)^2+(2x+3x^2)^3$$
$$\equiv 1+(6x+9x^2)+(12x^2+36x^3+27x^4)$$
$$+(8x^3+36x^4+54x^5+27x^6)$$
$$\equiv 1+6x+21x^2+44x^3+63x^4+54x^5+27x^6.$$

(vii) $$\left(x+\frac{2}{x}\right)^4+\left(2x-\frac{1}{x}\right)^4$$
$$\equiv \left(x^4+8x^2+24+\frac{32}{x^2}+\frac{16}{x^4}\right)+\left(16x^4-32x^2+24-\frac{8}{x^2}+\frac{1}{x^4}\right)$$
$$\equiv 17x^4-24x^2+48+\frac{24}{x^2}+\frac{17}{x^4}.$$

3. The binomial theorem; proof by induction. Assume that the binomial theorem is true for some particular positive integer k, so that

$$(x+a)^k = x^k + {}_kC_1 x^{k-1}a + \dots + {}_kC_p x^{k-p}a^p + \dots + {}_kC_k a^k.$$

Since $(x+a)^{k+1} = x(x+a)^k + a(x+a)^k$, an expression for $(x+a)^{k+1}$ is found by multiplying first by x, then by a, and adding (the first multiplication on the first line, the second on the second):

$$\begin{aligned} &x^{k+1} + {}_kC_1 x^k a + \dots + {}_kC_{p-1} x^{k-p+2} a^{p-1} \\ &\qquad\qquad + {}_kC_p \; x^{k-p+1} a^p + \dots + {}_kC_k \; xa^k \\ &\quad + x^k a + \dots + \dots \\ &\qquad\qquad + {}_kC_{p-1} x^{k-p+1} a^p + \dots + {}_kC_{k-1} xa^k + {}_kC_k a^{k+1}. \end{aligned}$$

The coefficient of the typical term $x^{k-p+1}a^p$ is

$${}_kC_p + {}_kC_{p-1}.$$

LEMMA. *To prove* (Vandermonde's theorem) *that*

$${}_kC_p + {}_kC_{p-1} = {}_{k+1}C_p.$$

By the formula for the combinations,

$$\begin{aligned} {}_kC_p + {}_kC_{p-1} &= \frac{k!}{p!\,(k-p)!} + \frac{k!}{(p-1)!(k-p+1)!} \\ &= \frac{k!\,\{(k-p+1)+p\}}{p!\,(k-p+1)!} \\ &= \frac{k!\,(k+1)}{p!\,(k-p+1)!} \\ &= \frac{(k+1)!}{p!\,(k+1-p)!} \\ &= {}_{k+1}C_p. \end{aligned}$$

Thus the typical term in the sum is

$${}_{k+1}C_p x^{k+1-p} a^p.$$

Moreover, for the second term of the summation,

$${}_kC_1 + 1 = k+1 = {}_{k+1}C_1,$$

and for the last term, $\quad {}_kC_k = 1 = {}_{k+1}C_{k+1}.$

Hence the assumption that the binomial theorem is true for $n = k$ leads to the result that

$$\begin{aligned} (x+a)^{k+1} = x^{k+1} + {}_{k+1}C_1 x^k a + \dots + {}_{k+1}C_p x^{k+1-p} a^p + \dots \\ + {}_{k+1}C_k xa^k + {}_{k+1}C_{k+1} a^{k+1}, \end{aligned}$$

which is the corresponding expansion for $k+1$.

But the result is true for $n = 1$, since

$$(x+a)^1 = x+a = x^1 + {}_1C_1 a^1.$$

Hence it is true for $n = 2$; and so for $n = 3$; and so, progressively, for every positive integral value of n.

Examples 2

Write down the following expansions:

1. $(x-a)^5$.
2. $(2x+a)^5$.
3. $(2x+3)^5$.
4. $(2x-5)^5$.
5. $(x+2a)^3$.
6. $(3x-5a)^4$.
7. $(1+a^2x)^6$.
8. $(ax-1)^7$.
9. $(ax-by)^3$.
10. $(abx-y^2)^5$.
11. $(1-abx)^6$.
12. $(x-1)^{12}$.
13. $(2x+1)^{10}$.
14. $(x-2y)^9$.
15. $(2x-3ay)^4$.
16. $(x^2+ay)^7$.

4. Approximations. The binomial theorem can be used to find approximate values for powers of numbers close to integers. The two Illustrations which follow indicate the process.

Illustration 2. To find an approximation to the value of $(2.03)^7$.

The expansion for $(2+x)^7$ is

$$2^7+7.2^6x+21.2^5x^2+35.2^4x^3+35.2^3x^4+21.2^2x^5+7.2x^6+x^7,$$

or

$$128+448x+672x^2+560x^3+280x^4+84x^5+14x^6+x^7.$$

For $x = 0{\cdot}03$, the addition gives:

$$\begin{aligned} 128 &= 128 \\ 448x &= 13{\cdot}44 \\ 672x^2 &= 0{\cdot}6048 \\ 560x^3 &= 0{\cdot}01512 \\ 280x^4 &= 0{\cdot}0000252 \\ 84x^5 &= 0{\cdot}0000020412 \\ 14x^6 &= 0{\cdot}000000010206 \\ x^7 &= 0{\cdot}00000000002187 \\ \hline (2{\cdot}03)^7 &= 142{\cdot}05994725142787 \end{aligned}$$

For most purposes, a calculation such as this is ridiculously exact, and an answer to a few places of decimals usually suffices. For example, suppose that an answer to 2 places of decimals is required. It becomes obvious on making the calculation that the terms involving x^4 and higher powers of x cannot affect the answer, and all that is wanted are the first three terms:

$$\begin{array}{r} 128 \\ 13\cdot44 \\ 0\cdot6048 \\ 0\cdot01512 \\ (0\cdot0000252) \\ \hline 142\cdot0599(4..) \end{array}$$

giving the approximation 142·06.

***Illustration* 3.** *To find an approximation, correct to 3 places of decimals, to the value of* $(0\cdot98)^6$.

Observe that $0\cdot98 = 1-0\cdot02$, and form the expansion

$$(1-x)^6 \equiv 1-6x+15x^2-20x^3+15x^4-6x^5+x^6.$$

Thus, with $x = 0\cdot02$,

$$\begin{aligned}(0\cdot98)^6 &= 1-6(0\cdot02)+15(0\cdot0004)-20(0\cdot000008)+\ldots \\ &= 1-0\cdot12+0\cdot006-0\cdot00016+\ldots \\ &= 1\cdot006-0\cdot12-\ldots \\ &= 0\cdot886\end{aligned}$$

to 3 places of decimals.

5. Properties of the coefficients.

(i) The derivation of the coefficients as combinations ensures that they are all positive integers. Those 'equidistant' from the two ends are equal.

(ii) The greatest binomial coefficient ${}_nC_p$ for given n may be found as follows:

It will be true that $${}_nC_{p-1} \leqslant {}_nC_p,$$

so long as $$\frac{n!}{(p-1)!\,(n-p+1)!} \leqslant \frac{n!}{p!\,(n-p)!},$$

or, on multiplying by the positive number $\frac{p!\,(n-p+1)!}{n!}$, so long as

$$p \leqslant n-p+1,$$

or $$p \leqslant \tfrac{1}{2}(n+1).$$

If n is an even integer, the greatest coefficient will be ${}_nC_p$, where

$$p = \tfrac{1}{2}(n+1)-\tfrac{1}{2} = \tfrac{1}{2}n.$$

If n is an odd integer, the greatest coefficient will be ${}_nC_p$, where

$$p = \tfrac{1}{2}(n+1).$$

The series has $n+1$ terms in all, so that *the binomial coefficients increase steadily till (n even) the middle term or (n odd) the two middle terms, and thereafter decrease steadily.*

For example, $n = 4$, coefficients 1, 4, 6, 4, 1;

$n = 5$, coefficients 1, 5, 10, 10, 5, 1.

(iii) The greatest numerical coefficient ${}_nC_pa^p$ for given n may be found by a similar device:

It will be true that $${}_nC_{p-1}a^{p-1} \leqslant {}_nC_pa^p,$$

so long as $$\frac{n!}{(p-1)!\,(n-p+1)!}a^{p-1} \leqslant \frac{n!}{p!\,(n-p)!}a^p.$$

If it is assumed that a is positive (or the *numerical value* of a taken otherwise) then

$$p \leqslant (n-p+1)a,$$

or $$p(1+a) \leqslant (n+1)a,$$

or $$p \leqslant \frac{a}{1+a}(n+1).$$

The greatest numerical coefficient ${}_nC_pa^p$ arises from the greatest positive integral value of p satisfying this inequality.

Illustration 4. *To find the greatest numerical coefficient in the expansion*

$$(x+3)^{17}.$$

It will be true that $${}_{17}C_{p-1}(3)^{p-1} \leqslant {}_{17}C_p(3)^p,$$

so long as $$\frac{17!}{(p-1)!\,(18-p)!}3^{p-1} \leqslant \frac{17!}{p!(17-p)!}3^p,$$

or $$p \leqslant 3(18-p),$$

or $$4p \leqslant 54.$$

The greatest value of p is 13, and so the greatest numerical coefficient is

$$\frac{17!\,3^{13}}{13!\,4!},$$

or 2380×3^{13}.

(iv) Certain summations can be effected by giving particular values to x and a. For example, with $x = 1$, $a = 1$,

$$2^n = 1+{}_nC_1+{}_nC_2+\ldots+{}_nC_p+\ldots+{}_nC_n;$$

and with $x = 1$, $a = -1$,

$$0 = 1-{}_nC_1+{}_nC_2-\ldots+(-1)^p{}_nC_p+\ldots+(-1)^n{}_nC_n.$$

Hence *the sum of the odd binomial coefficients is equal to the sum of the even binomial coefficients, each being* 2^{n-1}.

Thus the expansion

$$(x+a)^6 = x^6+6x^5a+15x^4a^2+20x^3a^3+15x^2a^4+6xa^5+a^6$$

gives the identities

$$1+15+15+1 = 2^5 = 32,$$
$$6+20+6 = 2^5 = 32.$$

6. Properties obtained by multiplication of series. The binomial expansion

$$(1+x)^n = 1+nx+\frac{n(n-1)}{2!}x^2+\ldots+x^n$$

may for convenience be written in the form

$$(1+x)^n = c_0+c_1x+c_2x^2+\ldots+c_nx^n,$$

where c_r is written for the coefficient

$${}_nC_r \equiv \frac{n!}{r!\,(n-r)!}.$$

The right-hand side is a polynomial in x of degree n.

Suppose, now, that we take two binomial expansions, say

$$(1+x)^2 = 1+2x+x^2,$$
$$(1+x)^n = c_0+c_1x+c_2x^2+\ldots+c_nx^n.$$

Their product is the binomial expression

$$(1+x)^{n+2},$$

whose expansion is

$$(1+x)^{n+2} = 1+(n+2)x+\frac{(n+2)(n+1)}{2!}x^2 +\frac{(n+2)(n+1)n}{3!}x^3+\ldots+x^{n+2}.$$

The product may be equated to this expansion, so there is an identity

$$(1+2x+x^2)(c_0+c_1x+c_2x^2+\ldots+c_nx^n)$$
$$\equiv 1+(n+2)x+\frac{(n+2)(n+1)}{2!}x^2+\ldots+x^n.$$

The left-hand side, when multiplied out, is a polynomial of degree $n+2$, which, being identical with the right-hand side, has equal coefficients with it (p. 70). Equating, for example, the coefficients of x, x^2, x^3, x^4, the following typical relations are obtained:

$$c_1+2c_0 \quad = n+2,$$

$$c_2+2c_1+c_0 = \frac{(n+2)(n+1)}{2},$$

$$c_3+2c_2+c_1 = \frac{(n+2)(n+1)n}{6},$$

$$c_4+2c_3+c_2 = \frac{(n+2)(n+1)n(n-1)}{24}.$$

More generally, the coefficient of x^r gives

$$c_r+2c_{r-1}+c_{r-2} = \frac{(n+2)(n+1)\ldots(\overline{n+2}-r+1)}{r!}$$
$$= \frac{(n+2)(n+1)\ldots(n+3-r)}{r!}.$$

This relation holds for $r = 2, 3, \ldots, n$.

Consider, again, the identity

$$(1+x)^n(1+x)^n \equiv (1+x)^{2n}.$$

On expansion, this is

$$(c_0+c_1x+c_2x^2+\ldots+c_nx^n)(c_0+c_1x+c_2x^2+\ldots c_nx^n)$$
$$\equiv 1+2nx+\frac{2n(2n-1)}{2!}x^2+\frac{2n(2n-1)(2n-2)}{3!}x^3+\ldots+x^{2n}.$$

Equating coefficients of, say, x^5 gives the identity

$$c_0c_5+c_1c_4+c_2c_3+c_3c_2+c_4c_1+c_5c_0 = \frac{2n(2n-1)(2n-2)(2n-3)(2n-4)}{5!},$$

or $$c_5+c_1c_4+c_2c_3 = \frac{n(2n-1)(2n-2)(2n-3)(2n-4)}{5!}.$$

Revision Examples IX

1. The coefficient of x^6 in the expansion of $(1+kx)^{10}$ in powers of x is given to be equal to the coefficient of x^8. Find the possible values of k.

2. Calculate the coefficient of x^4 in the expansion of $(1+2x-3x^2)^4$.

3. The expression

$$(1+\tfrac{1}{2}x)^{10}-(1+x)^5$$

is expanded with the help of the binomial theorem, and all terms in the result which contain the fifth and higher powers of x are neglected. Prove that the expression so obtained is

$$\tfrac{5}{8}(2x^2+8x^3+13x^4).$$

Hence, by putting $x = 0{\cdot}02$, calculate to five places of decimals the value of $(1{\cdot}01)^{10}-(1{\cdot}02)^5$.

4. Use the binomial theorem to write down the full expansion of $(1+x)^8$ in powers of x, and use your expansion to find the value of $(1{\cdot}01)^8$ correct to five significant figures.

5. You are given that, when $(x+a)^3(x-b)^6$ is expanded in powers of x, the coefficient of x^8 is 0 and the coefficient of x^7 is -9. Find all possible values of a and b.

6. By means of the binomial theorem, prove that

$$(2x+1)^5-(x+2)^5 = 31(x^5-1)+70(x^4-x)+40(x^3-x^2),$$

and deduce that, if $x > 1$, then

$$(2x+1)^5 > (x+2)^5.$$

7. Find the expansion of

$$(1+2x+3x^2)^4$$

in ascending powers of x, calculating the coefficients as far as the term in x^3.

8. Find the coefficient of x^5 in the expansion of

$$(1-x+2x^2)^5.$$

9. Use the binomial theorem to write down the expansion of $(1-x)^7$ in powers of x, and *hence* calculate the value of $(0{\cdot}98)^7$ to five significant figures.

10. It is known that x, a, n are positive integers and that the first three terms in the binomial expansion of $(x+a)^n$ are 729, 2916, 4860 respectively. Find x, a, n.

11. If y denotes $x+1/x$, express x^7+1/x^7 in the form

$$y^7+Ay^5+By^3+Cy,$$

where A, B, C are numerical coefficients.
[Consider $(x+1/x)^7$.]

12. If y denotes $x+1/x$, prove that

$$\left(x^7-\frac{1}{x^7}\right)\div\left(x-\frac{1}{x}\right)$$

can be expressed in the form $y^6+Ay^4+By^2+C$, and find the values of A, B, C.

13. Prove the identity

$$x^5+y^5 \equiv (x+y)^5-5xy(x+y)^3+5x^2y^2(x+y).$$

A positive number x and a negative number y exist such that

$$x+y = 1, \quad x^5+y^5 = 2101.$$

Use the above identity to form a quadratic in (xy), and deduce that $xy = -20$.
Hence find x and y.

14. Find the value of $(0{\cdot}998)^{40}$ correct to four places of decimals.

15. Find the term independent of x in the expansion of

$$\left(x+\frac{2}{x}\right)^{14}.$$

Which is the greatest term when $x = 3$?

16. Write down a formula for the coefficient of x^{2r} in the expansion of $(1-x^2)^n$ in powers of x, where n is a positive integer.
Find the coefficient of x^7 in the expansion of

$$(1-x)^8(1+x)^7.$$

17. Find the greatest term in the expansion of

$$(\tfrac{1}{3}+\tfrac{1}{2})^{13}.$$

18. Show that the term independent of x in the expansion of

$$\left(x-\frac{2}{x}\right)^{10}$$

is -8064.

19. Expand $(1+x)^5$ in ascending powers of x, and hence evaluate $(1{\cdot}01)^5$ to four significant figures.

20. Find the ratio of the seventh term to the sixth in the expansion of $(1+0{\cdot}05)^{13}$ by means of the binomial theorem.

Show that the sum of all the terms after the sixth is less than $0{\cdot}00003$.

21. Write down the expansion of $(a+b)^5$ and prove that, if

$$a+b+c = 0,$$

then $$5c^3a+10c^2a^2+10ca^3+5a^4 = 5c^3b+10c^2b^2+10cb^3+5b^4.$$

22. If c_r is the coefficient of x^r in the expansion of $(1+x)^n$, where n is a positive integer, prove that

$$c_0c_p-c_1c_{p-1}+c_2c_{p-2}-\ldots+(-1)^pc_pc_0$$

is equal to 0 if p is odd, and to $(-1)^{\frac{1}{2}p}c_{\frac{1}{2}p}$ if p is even.

23. If ${}_nc_r$ is the coefficient of x^r in the expansion of $(1+x)^n$, where n is a positive integer, prove that

$$\sum_{r=0}^{m} {}_{n-m}c_r \times {}_{n+m}c_{m-r} = {}_{2n}c_m.$$

24. Write down the expansions of

$$(\sqrt{a}+\sqrt{b})^5,\quad (\sqrt{a}-\sqrt{b})^5.$$

Prove that $$\left(\frac{101}{100}\right)^5-1-\frac{5}{100}-\frac{10}{100^2}$$

is less than $$\left(\frac{100}{99}\right)^5-1-\frac{5}{99}-\frac{10}{99^2}.$$

25. Show that

$$(a-b)^5 = a(a^2+10ab+5b^2)^2-b(5a^2+10ab+b^2)^2.$$

26. Find m, n so that the expansions, in ascending powers of x, of $(1-x^2)^{15}$ and of $(1+2x)^m(1-3x)^n$ may have the same coefficients of x and of x^2.

27. The expression $$(1+x+x^2)^4$$
is expanded in ascending powers of x. Prove that the first four terms are
$$1+4x+10x^2+16x^3.$$

28. When $x = (n^2+1)/2n$, prove that the $(r+1)$th term of the binomial expansion
$$(1+x)^n = 1+nx+\frac{n(n-1)}{2!}x^2+\ldots+x^n$$
(for positive integral n) is greater than the rth term provided that
$$(n+1)r < n^2+1,$$
and hence find which term of the expansion is greatest.

29. If the coefficient of x^r in the binomial expansion for $(1+x)^n$ is denoted by ${}_nc_r$, prove that, when $r \geqslant 2$,
$${}_nc_r+2\,{}_nc_{r-1}+{}_nc_{r-2} = {}_{n+2}c_r.$$

30. By considering a particular power of x in the binomial expansions of $(1+x)^3(1+x)^n$ and $(1+x)^{n+3}$, prove that
$$\frac{n(n-1)(n-2)(n-3)}{4!}+\frac{3n(n-1)(n-2)}{3!}+\frac{3n(n-1)}{2!}+n$$
$$= \frac{(n+3)(n+2)(n+1)n}{4!}.$$

31. Find the value of r for which the coefficients of x^{38-3r} and x^{35-3r} in the expansion of $(2x^2+[3/x])^{19}$ are equal.

32. The expansion of $(1+x)^{23}$ in a series of ascending powers of x is
$$1+c_1x+c_2x^2+\ldots+c_{23}x^{23}.$$
Find all the values of k for which
$$c_{k+1}+c_{k-1} \leqslant 2c_k.$$

33. Prove that, if
$$(1+x)^n \equiv c_0+c_1x+\ldots+c_nx^n,$$
then

(i) $c_0^2+c_1^2+c_2^2+\ldots+c_n^2 = \dfrac{(2n)!}{(n!)^2}.$

(ii) $c_0^2-c_1^2+c_2^2-\ldots\pm c_n^2 = 0 \quad (n \text{ odd})$
$$= \frac{(-1)^m(2m)!}{(m!)^2} \quad (n = 2m).$$

(iii) $c_0c_1+c_1c_2+c_2c_3+\ldots+c_{n-1}c_n = \dfrac{(2n)!}{(n+1)!(n-1)!}$.

[Consider the product of $(1+x)^n$ and $(1+x^{-1})^n$.]

(iv) $c_0c_2+c_1c_3+\ldots+c_{n-2}c_n = \dfrac{(2n)!}{(n+2)!(n-2)!}$.

(v) $c_0-c_1+c_2-c_3+\ldots+(-1)^r c_r = \dfrac{(-1)^r(n-1)!}{r!(n-r-1)!}$,

where $r < n$.

(vi) $1+2c_1+4c_2+8c_3+\ldots+2^n c_n = 3^n$.

(vii) $1-2c_1+4c_2-8c_3+\ldots+(-1)^n 2^n c_n = (-1)^n$.

15

THE SUMMATION OF SERIES

1. Definitions and notation. A set of numbers progressing in an orderly fashion is called a *sequence*. Typical examples are

$$1, 2, 3, 4, 5, 6, \ldots;$$

$$1, -\tfrac{1}{2}, \tfrac{1}{3}, -\tfrac{1}{4}, \tfrac{1}{5}, -\tfrac{1}{6}, \ldots;$$

$$x, x^3, x^5, x^7, x^9, x^{11}, \ldots;$$

$$(a+b), (a+2b)^2, (a+3b)^3, (a+4b)^4, (a+5b)^5, (a+6b)^6, \ldots;$$

$$1, -1, 1, -1, 1, -1, \ldots.$$

The term occupying the pth place in a sequence is called the pth term, and is often, for theoretical work of a general character, denoted by a symbol such as u_p, a_p, A_p. In the examples quoted above, the pth terms are respectively

$$p, \quad \frac{(-1)^{p-1}}{p}, \quad x^{2p-1}, \quad (a+pb)^p, \quad (-1)^{p-1}.$$

Warning Example. *What is the fifth term of the sequence*

$$1, 1, 1, 1, \ldots?$$

A *series* is formed by adding the terms of a sequence in order:

$$1+2+3+4+5+6+\ldots,$$

$$1-\tfrac{1}{2}+\tfrac{1}{3}-\tfrac{1}{4}+\tfrac{1}{5}-\tfrac{1}{6}+\ldots,$$

$$x+x^3+x^5+x^7+x^9+x^{11}+\ldots,$$

and so on. The process of summation is often denoted by the symbol Σ (sigma) placed in front of the pth term:

$$\Sigma p, \quad \Sigma\frac{(-1)^{p-1}}{p}, \quad \Sigma x^{2p-1},$$

and the range of terms to be summed is indicated by symbols such as

$$\sum_{p=1}^{8} p, \quad \sum_{p=1}^{15}\frac{(-1)^p}{p}, \quad \sum_{p=1}^{n} x^{2p-1},$$

meaning respectively the sum of the first 8 terms of the series Σp, the sum of the first 15 terms of the series $\Sigma \frac{(-1)^p}{p}$, the sum of the first n terms of the series Σx^{2p-1}. Again, a symbol such as

$$\sum_{p=5}^{9} p^2$$

would mean the summation

$$5^2+6^2+7^2+8^2+9^2,$$

and a symbol such as $\displaystyle\sum_{p=3}^{6} \frac{(-1)^p}{p^3}$

would mean the summation

$$-\tfrac{1}{27}+\tfrac{1}{64}-\tfrac{1}{125}+\tfrac{1}{216}.$$

When no doubt can arise, a symbol such as $\sum\limits_{p=1}^{n} p^5$ is often abbreviated to $\sum\limits_{1}^{n} p^5$, the summation being understood to be with respect to p.

Note that

$$\sum_{m}^{n} u_p = \sum_{1}^{n} u_p - \sum_{1}^{m-1} u_p.$$

The sum of the first n terms of a series is usually denoted by a symbol such as S_n; thus

$$S_n \equiv \sum_{p=1}^{n} u_p.$$

The following two summations are important:

(i) $\sum\limits_{p=1}^{n} 1 = n.$

(ii) $\sum\limits_{p=1}^{n} p = \tfrac{1}{2}n(n+1).$

To prove the latter, write

$$S_n = 1 \quad +2 \quad +\ldots+(n-1)+n,$$

so that $\quad S_n = n \quad +(n-1)+\ldots+2 \quad +1.$

Add: $\quad 2S_n = (n+1)+(n+1)+\ldots+(n+1)+(n+1)$

$$= n(n+1).$$

Hence $\quad S_n = \tfrac{1}{2}n(n+1).$

2. The arithmetic series. A sequence of numbers each of which exceeds its predecessor by a constant (positive or negative) amount is call an *arithmetic* sequence. Typical examples are

$$1,\ 4,\ 7,\ 10,\ 13,\ 16,\ 19,\ \ldots,$$

$$5,\ 3,\ 1,\ -1,\ -3,\ -5,\ -7,\ \ldots.$$

In general, if the first time is a and the constant common difference (as it is called) is d, then the sequence is

$$a,\ a+d,\ a+2d,\ a+3d,\ a+4d,\ \ldots.$$

The pth term is $$a+(p-1)d.$$

If the sum of the first n terms of the corresponding series is S_n, then

$$S_n = \sum_1^n \{a+(p-1)d\}$$

$$= a\sum_1^n 1 + d\sum_1^n (p-1)$$

$$= an + \tfrac{1}{2}d(n-1)n,$$

by the formulae on p. 180.
Hence we have the formula for the sum of the first n terms:

$$S_n = \tfrac{1}{2}n\{2a+(n-1)d\}.$$

An alternative form which is sometimes convenient is

$$S_n = \tfrac{1}{2}n\{\text{first term} + \text{last term}\}.$$

Illustration 1. *To find the sum of the first n odd numbers.*

The odd numbers form the arithmetic series

$$1+3+5+7+\ldots$$

in which $a = 1$, $d = 2$. Thus

$$\begin{aligned} S_n &= \tfrac{1}{2}n\{2a+(n-1)d\} \\ &= \tfrac{1}{2}n\{2+(n-1)2\} \\ &= \tfrac{1}{2}n(2n) \\ &= n^2. \end{aligned}$$

Illustration 2. *To find what number of terms of the arithmetic series*

$$1+4+7+10+13+\ldots$$

must be taken if the sum is to exceed 100.

The series is an arithmetic one, in which $a = 1$, $d = 3$. The sum of the first n terms is thus

$$\begin{aligned} &\tfrac{1}{2}n\{2a+(n-1)d\} \\ &= \tfrac{1}{2}n\{2+(n-1)3\} \\ &= \tfrac{1}{2}n(3n-1). \end{aligned}$$

Hence the relation for a sum exceeding 100 is

$$\tfrac{1}{2}n(3n-1) > 100,$$

or $$n(3n-1) > 200,$$

or
$$3n^2-n-200 > 0,$$
or
$$(3n-25)(n+8) > 0.$$

Now n is necessarily positive, so that, for relevant values, $n+8$ is positive. Hence the condition is
$$3n-25 > 0,$$
or
$$n > 8\tfrac{1}{3}.$$
Hence the number of terms to be taken is 9. The actual sum for 9 terms is
$$\tfrac{9}{2}(27-1) = \tfrac{9}{2}\times 26,$$
or 117.

***Illustration* 3.** *To find the first term and the common difference of an arithmetic series in which the sum of the first* 11 *terms is* 352 *and the sum of the next* 10 *terms is* 845.

In the standard notation, the two given facts are:
$$\tfrac{11}{2}(2a+10d) = 352,$$
$$\tfrac{21}{2}(2a+20d) = 352+845 = 1197.$$
Thus
$$2a+10d = 64,$$
$$2a+20d = 114.$$
Subtract:
$$10d = 50,$$
so that
$$d = 5.$$
Hence
$$a = 7.$$

3. Generalized arithmetic series. The simplest arithmetic series is
$$1+2+3+4+5+\ldots, \qquad \text{(p. 180)}$$
and may be used as a basis to explain some extensions. The generalized arithmetic series to be considered are
$$1.2+2.3+3.4+4.5+5.6+\ldots,$$
$$1.2.3+2.3.4+3.4.5+4.5.6+5.6.7+\ldots,$$
$$1.2.3.4+2.3.4.5+3.4.5.6+4.5.6.7+5.6.7.8+\ldots,$$
and so on. The pth terms are
$$p(p+1),$$
$$p(p+1)(p+2),$$
$$p(p+1)(p+2)(p+3).$$

These series are summed by a device which may be illustrated by reference to the first and third. The aim is to express successive terms as *differences*, leading to a summation of the type
$$(a-b)+(b-c)+(c-d)+\ldots+(m-n) = a-n.$$

(i) *To sum the series*

$$S_n \equiv \sum_{p=1}^{n} p(p+1).$$

Define a new sequence, whose pth term u_p is given by the formula

$$u_p = p(p+1)(p+2).$$

Then
$$u_{p-1} = (p-1)\,p(p+1),$$

so that
$$u_p - u_{p-1} = p(p+1)\{(p+2)-(p-1)\}$$
$$= 3p(p+1).$$

Hence
$$p(p+1) = \tfrac{1}{3}(u_p - u_{p-1}).$$

The given series thus satisfies the relation

$$S_n = \tfrac{1}{3}(u_1-u_0)+\tfrac{1}{3}(u_2-u_1)+\tfrac{1}{3}(u_3-u_2)+\ldots+\tfrac{1}{3}(u_n-u_{n-1})$$

[where $u_0 = 0$, so that the formula '$1.2 = \tfrac{1}{3}(u_1-u_0)$' is true for the first term also]. Hence, removing brackets and cancelling,

$$S_n = \tfrac{1}{3}(-u_0+u_n)$$
$$= \tfrac{1}{3}u_n,$$

so that
$$S_n = \tfrac{1}{3}n(n+1)(n+2).$$

(ii) *To sum the series*

$$S_n \equiv \sum_{p=1}^{n} p(p+1)(p+2)(p+3).$$

Define a new sequence whose pth term is given by the formula

$$u_p = p(p+1)(p+2)(p+3)(p+4).$$

Then
$$u_{p-1} = (p-1)\,p(p+1)(p+2)(p+3),$$

so that
$$u_p - u_{p-1} = p(p+1)(p+2)(p+3)\{(p+4)-(p-1)\}$$
$$= 5p(p+1)(p+2)(p+3).$$

Hence
$$p(p+1)(p+2)(p+3) = \tfrac{1}{5}(u_p - u_{p-1}).$$

The given series thus satisfies the relation

$$S_n = \tfrac{1}{5}(u_1-u_0)+\tfrac{1}{5}(u_2-u_1)+\tfrac{1}{5}(u_3-u_2)+\ldots+\tfrac{1}{5}(u_n-u_{n-1})$$
$$= \tfrac{1}{5}(u_n-u_0),$$

where $u_0 = 0$. Hence

$$S_n = \tfrac{1}{5}u_n$$
$$= \tfrac{1}{5}n(n+1)(n+2)(n+3)(n+4).$$

More generally, it may be proved that

$$1+2+3+\ldots+n = \tfrac{1}{2}n(n+1),$$
$$1.2+2.3+3.4+\ldots+n(n+1) = \tfrac{1}{3}n(n+1)(n+2),$$
$$1.2.3+2.3.4+3.4.5+\ldots+n(n+1)(n+2) = \tfrac{1}{4}n(n+1)(n+2)(n+3),$$

and so on.

Illustration **4.** *To find the sum of the squares of the first n positive integers.*

The pth term of the sequence is p^2, which may be written in the form

$$p(p+1)-p,$$

so that the sum is
$$\sum_1^n p(p+1)-\sum_1^n p$$
$$= \tfrac{1}{3}n(n+1)(n+2)-\tfrac{1}{2}n(n+1)$$
$$= \tfrac{1}{6}n(n+1)\{2(n+2)-3\}$$
$$= \tfrac{1}{6}n(n+1)(2n+1).$$

Illustration **5.** *To find the sum of the cubes of the first n positive integers.*

The pth term of the sequence is p^3, which may be written in the form

$$p(p+1)(p+2)-3p^2-2p \equiv p(p+1)(p+2)-3p(p+1)+p,$$

so that the sum is

$$\sum_1^n p(p+1)(p+2)-3\sum_1^n p(p+1)+\sum_1^n p$$
$$= \tfrac{1}{4}n(n+1)(n+2)(n+3)-n(n+1)(n+2)+\tfrac{1}{2}n(n+1)$$
$$= \tfrac{1}{4}n(n+1)\{(n^2+5n+6)-4(n+2)+2\}$$
$$= \tfrac{1}{4}n(n+1)(n^2+n)$$
$$= \tfrac{1}{4}n^2(n+1)^2$$
$$= \{\tfrac{1}{2}n(n+1)\}^2.$$

Illustration **6.** *To find the number of 'pips' in a set of dominoes.* Each domino is in two parts, each of which may contain independently any number of 'pips' from 0 to 6.

Consider, then, the $r+1$ dominoes

$$(0, r), (1, r), (2, r), \ldots, (r, r)$$

for $r \leqslant 6$. The total number of 'pips' is

$$0+1+2+\ldots+r+(r+1)r$$
$$= \tfrac{1}{2}r(r+1)+(r+1)r$$
$$= \tfrac{3}{2}r(r+1).$$

The grand total is thus

$$\sum_{r=0}^{6} \tfrac{3}{2}r(r+1) \equiv \sum_{r=1}^{6} \tfrac{3}{2}r(r+1)$$

$$= \frac{3}{2} \cdot \frac{6.7.8}{3}$$

$$= 168.$$

4. The geometric series. A sequence of numbers, each of which bears to its predecessor a constant (positive or negative) ratio is called a *geometric* sequence. Typical examples are

$$1, 2, 4, 8, 16, 32, 64, \ldots,$$

$$1, -\tfrac{1}{3}, \tfrac{1}{9}, -\tfrac{1}{27}, \tfrac{1}{81}, -\tfrac{1}{243}, \tfrac{1}{729}, \ldots.$$

In general, if the first term is a and the constant common ratio (as it is called) is r then the sequence is

$$a, ar, ar^2, ar^3, ar^4, \ldots.$$

The pth term is $$ar^{p-1}.$$

If the sum of the first n terms of the series having these terms is S_n, then

$$S_n \equiv a + ar + ar^2 + \ldots + ar^{n-1},$$

the last term being ar^{n-1}.

A formula for this sum may be obtained by the device of multiplying it throughout by r:

$$rS_n \equiv ar + ar^2 + ar^3 + \ldots + ar^n.$$

Subtracting these two expressions and noting that the terms $ar, ar^2, \ldots, ar^{n-1}$ then cancel, we have the relation

$$(1-r)S_n = a - ar^n = a(1-r^n).$$

Hence $$S_n = \frac{a(1-r^n)}{1-r}.$$

There is an implication in this way of writing the formula that r is less than 1; if r is greater than 1, we may, if we wish, use the equivalent form

$$S_n = \frac{a(r^n-1)}{r-1}.$$

It is assumed that r is not equal to 1. The sum then would just be na.

Illustration 7. *To find the sum of the first n powers of* 2.

The powers of 2 form the geometric series

$$1+2+4+8+16+\dots,$$

in which $a = 1$, $k = 2$. Thus

$$S_n = \frac{a(r^n-1)}{r-1}$$

$$= \frac{1(2^n-1)}{2-1}$$

$$= 2^n-1.$$

Illustration 8. *To find how many terms of the series*

$$1+\tfrac{1}{2}+\tfrac{1}{4}+\tfrac{1}{8}+\tfrac{1}{16}+\dots$$

must be taken so that the difference between the sum and 2 *is less than* $\frac{1}{1000}$.

Since $a = 1$, $r = \frac{1}{2}$, the sum of the first n terms is S_n, where

$$S_n = \frac{1\{1-(\frac{1}{2})^n\}}{1-\frac{1}{2}}$$

$$= 2\{1-(\tfrac{1}{2})^n\}$$

$$= 2-1/2^{n-1}.$$

Thus $$2-S_n = 1/2^{n-1}.$$

The larger n is taken, the smaller does $1/2^{n-1}$ become, and so the more nearly does S_n approach 2. In particular, if $2-S_n$ is less than $\frac{1}{1000}$ then

$$1/2^{n-1} \text{ is less than } \tfrac{1}{1000},$$

or $$2^{n-1} \text{ is greater than } 1000.$$

The first power of 2 to exceed 1000 is $2^{10} \equiv 1024$, so that n is given by the relation $n-1 = 10$, or $n = 11$.

5. The harmonic series, and generalizations. A series of numbers whose terms are the reciprocals of corresponding members of an arithmetic series is called a *harmonic* series. Typical series are

$$1+\tfrac{1}{3}+\tfrac{1}{5}+\tfrac{1}{7}+\tfrac{1}{9}+\dots,$$

$$\tfrac{1}{2}+\tfrac{1}{7}+\tfrac{1}{12}+\tfrac{1}{17}+\tfrac{1}{22}+\dots.$$

There is no formula corresponding to $\frac{1}{2}n\{2a+(n-1)d\}$ for an arithmetic series and $a(1-r^n)/(1-r)$ for a geometric series to express the sum of the first n terms of a harmonic series.

On the other hand, the *generalized harmonic series*, corresponding to the generalized arithmetic series (p. 182) can be summed readily:

(i) *To sum the series*

$$S_n \equiv \sum_{p=1}^{n} \frac{1}{p(p+1)}$$

$$\equiv \frac{1}{1.2}+\frac{1}{2.3}+\ldots+\frac{1}{n(n+1)}.$$

Define a new sequence, whose pth term v_p is given by the formula

$$v_p = \frac{1}{p}.$$

Then $$v_{p+1} = \frac{1}{p+1},$$

so that $$v_p - v_{p+1} = \frac{1}{p}-\frac{1}{p+1} = \frac{1}{p(p+1)}.$$

The given series thus satisfies the relation

$$S_n = (v_1-v_2)+(v_2-v_3)+\ldots+(v_n-v_{n+1})$$

$$= v_1 - v_{n+1}$$

$$= 1-\frac{1}{n+1}$$

$$= \frac{n}{n+1}.$$

NOTE. This method extends readily to other similar series, as we shall see. In this particular case it is simpler to observe at once that

$$\frac{1}{p}-\frac{1}{p+1} = \frac{1}{p(p+1)},$$

so that $$S_n = (1-\tfrac{1}{2})+(\tfrac{1}{2}-\tfrac{1}{3})+(\tfrac{1}{3}-\tfrac{1}{4})+\ldots+\left(\frac{1}{n}-\frac{1}{n+1}\right)$$

$$= 1-\frac{1}{n+1}.$$

(ii) *To sum the series*

$$S_n = \sum_{p=1}^{n} \frac{1}{p(p+1)(p+2)(p+3)}$$

$$= \frac{1}{1.2.3.4}+\frac{1}{2.3.4.5}+\ldots+\frac{1}{n(n+1)(n+2)(n+3)}.$$

Define a new sequence, whose pth term v_p is given by the formula

$$v_p = \frac{1}{p(p+1)(p+2)}.$$

Then
$$v_{p+1} = \frac{1}{(p+1)(p+2)(p+3)},$$
so that
$$v_p - v_{p+1} = \frac{(p+3)-p}{p(p+1)(p+2)(p+3)} = \frac{3}{p(p+1)(p+2)(p+3)}.$$
The given series thus satisfies the relation
$$\begin{aligned} S_n &= \tfrac{1}{3}\{(v_1-v_2)+(v_2-v_3)+\ldots+(v_n-v_{n+1})\} \\ &= \tfrac{1}{3}(v_1-v_{n+1}) \\ &= \frac{1}{3}\left\{\frac{1}{1.2.3.} - \frac{1}{(n+1)(n+2)(n+3)}\right\} \\ &= \frac{1}{18} - \frac{1}{3(n+1)(n+2)(n+3)}. \end{aligned}$$

6. Arithmetic, geometric and harmonic means. When two numbers a, b are given, it is possible to insert any given number of terms between them so that the resulting sequence may be arithmetic, geometric or harmonic. Suppose that k numbers are to be inserted.

(i) ARITHMETIC. There are $k+2$ terms, the first being a and the last b. The difference d is thus given by the formula
$$b = a+(k+1)d,$$
or
$$d = \frac{b-a}{k+1}.$$

(ii) GEOMETRIC. The ratio r is given by the formula
$$b = ar^{k+1},$$
or
$$r = \sqrt[k+1]{\left(\frac{b}{a}\right)}.$$

When k is odd, b and a must have opposite signs for the existence of an even root; there are two solutions, equal in value but opposite in sign.

(iii) HARMONIC. The inverses form an arithmetic series with first term $1/a$ and last term $1/b$. Suppose that the difference is d. Then
$$\frac{1}{b} = \frac{1}{a}+(k+1)d,$$
so that
$$d = \frac{a-b}{(k+1)ab}.$$
When this arithmetic series has been found, the reciprocals give the harmonic series.

In particular, when $k = 1$,

(*a*) The ARITHMETIC MEAN between a, b is

$$\tfrac{1}{2}(a+b).$$

(*b*) The GEOMETRIC MEAN between a, b is

$$\sqrt{(ab)},$$

where ab is assumed to be positive. (The sign of the square root is usually taken to be that of a and b.)

(*c*) The HARMONIC MEAN between a, b is

$$\frac{2ab}{a+b}$$

(since $\frac{1}{a}, \frac{1}{2a}+\frac{1}{2b}, \frac{1}{b}$ are in arithmetic progression).

***Illustration* 9.** (A 'mixed' arithmetic and geometric series.) *To find the sum of the first n terms of the series*

$$1+3x+5x^2+\ldots+(2r-1)x^{r-1}+\ldots.$$

Let the sum be S, where

$$S = 1+3x+5x^2+\ldots+(2r-1)x^{r-1}+\ldots+(2n-1)x^{n-1}.$$

Then

$$xS = \qquad x+3x^2+\ldots+(2r-3)x^{r-1}+\ldots+(2n-3)x^{n-1}+(2n-1)x^n.$$

Subtract:

$$\begin{aligned}(1-x)S &= 1+2x+2x^2+\ldots+2x^{r-1}+\ldots+2x^{n-1}-(2n-1)x^n\\ &= 1-(2n-1)x^n+2(x+x^2+\ldots+x^{n-1})\\ &= 1-(2n-1)x^n+\frac{2x(1-x^{n-1})}{1-x}.\end{aligned}$$

Hence
$$S = \frac{1-(2n-1)x^n}{1-x}+\frac{2x(1-x^{n-1})}{(1-x)^2}.$$

Revision Examples X

1. How many terms must be taken of the arithmetical series

$$4+4\tfrac{1}{2}+5+\ldots$$

to give a sum of 250?

2. Find the sum of all the positive terms of the arithmetic series

$$35+33\tfrac{1}{2}+32+\ldots.$$

3. The eighth term of an arithmetic series is twice the third term; the sum of the first eight terms is 39. Find the first three terms, and show that the sum of the first n terms is $\frac{3}{8}n(n+5)$.

4. Find which term of the arithmetic series

$$8+7\tfrac{2}{3}+7\tfrac{1}{3}+\ldots$$

is the first negative term.

Find also the sum of the first n terms. Determine the smallest value of n for which this sum is negative, and calculate the corresponding value of the sum.

5. Find the sum of all the integers between 300 and 500 which are exactly divisible by 7.

6. The first term of an arithmetic series is 3, and the sum of the first three terms is equal to the sum of the first six terms. Prove that the sum of the first n terms is $\frac{3}{8}n(9-n)$.

7. Find the sum of the first n terms of the arithmetic series

$$2+5+8+\ldots,$$

and find the value of n for which the sum of $2n$ terms exceeds the sum of n terms by 224.

8. You are given that, for all values of n, the sum of n terms of a certain series is $2n^2-3n$. Find the first four terms of the series, and prove that they are the terms of an *arithmetic* series.

9. Find which term of the arithmetic series

$$19+18\tfrac{1}{3}+17\tfrac{2}{3}+\ldots$$

is the first negative term.

Find also the sum of the first n terms, and determine the smallest value of n for which the sum is negative.

10. Find the sum of all the odd numbers between 900 and 1000.

Find the sum of all the odd numbers between 900 and 1000 which are exactly divisible by 3.

11. Given that a and c are the first and third terms of an arithmetic series, find the common difference and the sum of the first nine terms.

12. Sum the arithmetic series $3+5+7+9+\ldots$ to 200 terms.

13. Find the sum of the first 100 terms of the arithmetic series whose first and second terms are 2 and $2\frac{1}{4}$.

Find the least number of terms of the geometric series whose first and second terms are 2 and $2\frac{1}{4}$ which must be taken so that their sum may exceed the sum in the first part of this question.

14. Find the sum of all even numbers from 4 to 100 inclusive, excluding those which are multiples of 3.

15. The ninth term of an arithmetic series is -1, and the sum of the first nine terms is 45. Find the common difference and the sum of the first 20 terms.

16. Prove that the sum of the first n terms of the series

$$\log a + \log ax + \log ax^2 + \ldots,$$

whose rth term is $\log ax^{r-1}$, is

$$n \log a + \tfrac{1}{2}n(n-1)\log x.$$

17. I lay aside every year £20 more than I laid aside the year before. I laid aside £100 the first year. How many years will it take me to lay aside £5800?

18. Prove that the sum of the odd numbers from 1 to 55 inclusive is equal to the sum of the odd numbers from 91 to 105 inclusive.

19. The houses of a row are numbered consecutively from 1 to 49. Show that there is a value of x such that the sum of the numbers of the houses before the house numbered x is equal to the sum of the houses after it, and find this value of x.

20. In a certain year a man saves £100 and in each succeeding year he saves £5 more than in the previous year. In how many years will his total savings (not including any interest they may have earned) first exceed £3000?

21. Prove that

$$\log x + \log(xy) + \log(xy^2) + \ldots$$

is an arithmetic series, and show, without using tables, that, when $x = 160$, $y = \frac{1}{2}$, the sum of the first 9 terms of the progression (the logarithms being to the base 10) is 9.

22. A and B begin work together. A's initial salary is £200 a year and he has an annual increment of £20. B is paid at first at the rate of £80 a year and has an increment of £8 every half-year. At the end of how many years will B have received more money than A?

23. If the mth and nth terms of an arithmetic series are in the ratio $2m-1:2n-1$, prove that the sum of the first m terms is to the sum of the first n terms as m^2 is to n^2.

24. Calculate the sum of all numbers between 0 and 201 which are multiples of 5 or 7; that is, find the sum of the series

$$5+7+10+14+15+...+35+...+200.$$

25. Given that the tenth term of an arithmetic series is 32, and that the sum of the first twenty terms is 670, find the first term and the common difference.

What is the least number of terms that must be taken if the sum is to exceed 2000?

26. Find the values of

$$1-2+3-4+... \quad \text{to } 2n \text{ terms,}$$
$$1-2+3-4+... \quad \text{to } 2n+1 \text{ terms,}$$

where n is a positive integer.

27. Find the sum of the first $3n$ terms of the series

$$1+3-5+7+9-11+13+15-17+....$$

28. Find how many terms of the series

$$2-3+4-5+...$$

are required to make the sum of the series equal to 35.

29. The roots of the equation

$$16x^4-64x^3+56x^2+16x-15 = 0$$

are known to be in arithmetical progression. Find them.

30. The ninety-first term of an arithmetical progression is 277, and the sum of the first eighty terms is 10,040. Prove that the eighth term is four times the first term.

31. Find the sum of the integers which lie between 1234 and 2345 inclusive.

How many numbers which are not multiples of 2 or of 3 lie between 1234 and 2345?

32. The formula for the sum of n terms of a certain geometric series is known to be 3^n-1 for all values of n. Find the first three terms.

33. Show that there are two possible geometric series in each of which the first term is 8, and the sum of the first three terms is 14. Find the second term and the sum of the first seven terms of each series.

34. Find the least number of terms of the geometric series

$$1+3+9+27+\ldots$$

which must be taken for the sum to exceed a thousand million.

35. Prove that the sum of the first twenty terms of the geometric series

$$7+2{\cdot}1+0{\cdot}63+\ldots$$

differs from 10 by an amount which is less than half of the twentieth term.

36. Given that 27 is the fourth term of a geometric series whose first term is 8, find the common ratio and the seventh term.

37. Sum the geometric series $2+6+18+54+\ldots$ to 30 terms, and verify, by using logarithms, that this sum is greater than 2×10^{14}.

38. Find the sum of the series

$$\frac{a}{k^6}+\frac{a}{k^5}+\ldots+\frac{a}{k}+a+ak+\ldots+ak^6.$$

39. Prove that the sum of the first $2m$ terms of the series

$$x+x^2+x^4+x^5+x^7+x^8+\ldots+x^{3r+1}+x^{3r+2}+\ldots$$

is

$$\frac{(x+x^2)(1-x^{3m})}{1-x^3}.$$

40. Find the least value of n for which

$$1+2+2^2+\ldots+2^{n-1}$$

exceeds 10,000.

41. Find the conditions that the roots of the equation

$$x^3+3px^2+3qx+r = 0$$

should be (i) in arithmetic progression, (ii) in geometric progression, (iii) in harmonic progression.

42. The equation

$$x^5+5a_4x^4+10a_3x^3+10a_2x^2+5a_1x+a_0 = 0$$

has three equal roots each equal to the arithmetic mean of the other two. Prove that

$$a_0 = 10a_3a_4^3-9a_4^5.$$

43. Solve the equation

$$81x^4+54x^3-189x^2-66x+40 = 0,$$

given that the roots are in arithmetic progression.

44. Prove that

$$1.3+3.5+5.7+\ldots+(2n-1)(2n+1) = \tfrac{1}{3}n(4n^2+6n-1).$$

45. Find the sum of the first n terms of the series

$$x+2x^2+3x^3+\ldots.$$

46. Find the sum of the first n terms of the series

$$1.3.5+3.5.7+5.7.9+\ldots.$$

47. Find the value of n such that

$$\sum_{r=n+1}^{2n}(r^2-48r) = 2\sum_{r=1}^{n}(r^2-48r).$$

48. Prove that, if n is even, the sum of the first n terms of the series

$$1^2+2.2^2+3^2+2.4^2+5^2+2.6^2+\ldots$$

is $\tfrac{1}{2}n(n+1)^2$, and find the sum if n is odd.

49. Find the sum of the first n terms of the series

(i) $$\frac{1}{1.3}+\frac{1}{2.4}+\frac{1}{3.5}+\frac{1}{4.6}+\ldots;$$

(ii) $$\frac{1}{1.4}+\frac{1}{4.7}+\frac{1}{7.10}+\frac{1}{10.13}+\ldots.$$

50. Find the sum of the first n terms of the series

$$1+3x+5x^2+7x^3+\ldots.$$

51. Find the sum of the first n terms of the series

$$1+4+9+16+\ldots.$$

Find the value of n for which 7 times the sum of the first $2n$ terms is 50 times the sum of the first n terms.

52. Prove that the sum of the first n terms of the series whose rth term is $r(r+1)(2r+1)$ is

$$\tfrac{1}{2}n(n+1)^2(n+2).$$

53. Prove that the sum of the first n terms of the series whose rth term is $r(r+2)(r+4)$ is

$$\tfrac{1}{4}n(n+1)(n+4)(n+5).$$

54. Prove that the sum of the first n terms of the series whose rth term is $(2r-1)^3$ is

$$n^2(2n^2-1).$$

55. Prove by induction that the sum of the first n terms of the series

$$\frac{1}{2}+\frac{1.3}{2.4}+\frac{1.3.5}{2.4.6}+\frac{1.3.5.7}{2.4.6.8}+\dots$$

is

$$\frac{1.3.5.\ \dots\ (2n+1)}{2.4.6.\ \dots\ 2n}-1.$$

56. Find the sum of the first n terms of the series

$$1.2.5+2.3.6+3.4.7+\dots.$$

57. Prove that

$$\frac{1.2+3.4+5.6+\dots+79.80}{2.3+4.5+6.7+\dots+80.81}=\frac{53}{55}.$$

58. Express

$$\frac{6(x+10)}{x(x+3)(x+5)}$$

in partial fractions.

Hence find the sum of the first n terms of the series

$$\frac{11}{1.4.6}+\frac{12}{2.5.7}+\frac{13}{3.6.8}+\frac{14}{4.7.9}+\dots.$$

59. Find the sum of the first n terms of the series

$$\frac{1}{2.5.8}+\frac{1}{5.8.11}+\frac{1}{8.11.14}+\dots.$$

60. Sum to n terms the series whose rth terms are

(i) $r(r+2)$;

(ii) $(r+2)x^r$.

61. Find the sum to n terms of the series whose rth term is

$$\frac{1}{1+2+3+\dots+r}.$$

62. Find the sum to n terms of the series

$$\frac{4}{1.2.3}+\frac{5}{2.3.4}+\frac{6}{3.4.5}+\dots.$$

63. Sum to n terms the series

$$\frac{1}{1.6}+\frac{1}{6.11}+\frac{1}{11.16}+\dots.$$

64. Sum to n terms the series

$$\frac{1.2.12}{4.5.6}+\frac{2.3.13}{5.6.7}+\frac{3.4.14}{6.7.8}+\dots.$$

65. Evaluate $\quad 1^3-3^3+5^3-7^3+\dots+37^3-39^3.$

66. Sum to n terms the series

(i) $\dfrac{1}{1.3.5}+\dfrac{2}{3.5.7}+\dfrac{3}{5.7.9}+\dots;$

(ii) $\dfrac{1}{3.7.11}+\dfrac{5}{7.11.15}+\dfrac{9}{11.15.19}+\dots;$

67. Prove that, if r is a positive integer,

$$\frac{1}{r^2}>\frac{1}{r(r+1)},$$

and deduce that $\quad \displaystyle\sum_{r=m+1}^{2m}\frac{1}{r^2}>\frac{m}{(m+1)(2m+1)}.$

16

INFINITE SERIES

The ideas of limits, convergence, infinite series, and similar topics may be approached through algebra, through calculus, or, preferably, through both in partnership. For the approach through calculus, see, for example, the author's *An Analytical Calculus*. The present treatment is more algebraic in bias, and aims at rendering reasonable certain results (general binomial series, exponential series, logarithmic series) which the reader may wish to use before he reaches the stage of being able to follow the full rigours of a detailed proof. It is hoped, however, that the existence of gaps in the argument is made sufficiently clear when necessary. Some of the steps are admittedly difficult, and the beginner may wish to walk gently at times.

1. Limits. Consider the sequence

$$\begin{aligned}&0{\cdot}9,\\&0{\cdot}99,\\&0{\cdot}999,\\&0{\cdot}9999,\\&\dots\dots\dots\end{aligned}$$

Its elements are the sums $S_1, S_2, S_3, S_4, \ldots$ of the first 1, 2, 3, 4, ... terms of the geometric series

$$\frac{9}{10}+\frac{9}{10^2}+\frac{9}{10^3}+\frac{9}{10^4}+\ldots,$$

so that (p. 185)
$$S_n = \frac{\frac{9}{10}\left(1-\frac{1}{10^n}\right)}{1-\frac{1}{10}} = \frac{\frac{9}{10}\left(1-\frac{1}{10^n}\right)}{\frac{9}{10}},$$

$$= 1-\frac{1}{10^n}.$$

Hence
$$1-S_n = \frac{1}{10^n},$$

and so, for each value of n, the sum S_n falls short of 1 by the fraction $1/10^n$, as, indeed, is clear intuitively.

Now, and this is the crux of all that follows, we can ensure that this difference $1/10^n$ is as small as we please by taking n sufficiently large; the decimal 0·9999 ... can be made to approximate as closely to 1 as we please by taking a sufficient number of places.

Suppose that we agree to fix on some pre-determined test-number, which we shall call a *gauge*, chosen so small that anything less than it may be regarded as negligible. The actual value of the gauge may vary, as required, from problem to problem; the important thing is that it can be settled before the calculation starts, so that we know *from the beginning* how fine it is to be.

In the present instance, we might have set the gauge at $\frac{1}{1\,000\,000}$; by taking $n = 7$ we should have got a value 0·9999999 which came closer to the value 1 than the gauge required. We might have set it at $\frac{1}{1\,000\,000\,000}$; by taking $n = 10$ we should have got a value 0·9999999999 which came closer to the value 1 than the gauge required. However fine we made the gauge, we should still have been able to produce a decimal to come closer to the value 1 than the gauge required.

DEFINITION. *When a sequence* $S_1, S_2, S_3, S_4, \ldots$ (like that just considered) *is such that all its members, after a certain ascertainable one, come closer to a number l than any gauge number we care to select in advance (however small that gauge may be), then we say that the sequence* $S_1, S_2, S_3, S_4, \ldots$ *tends to the limit l as n tends to infinity.*

The word 'infinity' is often replaced by the symbol ∞.

We write this in the form

$$S_n \to l \quad \text{as} \quad n \to \infty,$$

or

$$\lim_{n\to\infty} S_n = l,$$

where it is implicit in this context that n is always a positive integer.

For example, if

$$S_n \equiv 0{\cdot}999 \ldots 9,$$

to n places of decimals, then

$$S_n \to 1 \quad \text{as} \quad n \to \infty,$$

or

$$\lim_{n\to\infty} S_n = 1.$$

Another example, which occurred incidentally in the work just given, is that, if

$$S_n = \frac{1}{10^n},$$

then

$$S_n \to 0 \quad \text{as} \quad n \to \infty.$$

A series for which the sum S_n of the first n terms tends to a limit S is said to *converge*, or to be *convergent*.

It will often be convenient to denote the gauge number by the letter g, which will be regarded as small, and necessarily positive. It is sometimes handier to express it in the alternative form $1/G$, where G is regarded as large, and necessarily positive.

Illustration 1. *The sequence* $(3n+2)/(2n+1)$. The quotient $(3n+2)/(2n+1)$ takes, for $n = 1, 2, 3, \ldots$ in succession, the sequence of values $S_1, S_2, S_3, \ldots,$ where

$$S_1 = \tfrac{5}{3},\quad S_2 = \tfrac{8}{5},\quad S_3 = \tfrac{11}{7},\quad S_4 = \tfrac{14}{9},\quad \ldots.$$

We prove that *this sequence tends to the limit* $\tfrac{3}{2}$ *as n tends to infinity.*

Let the gauge number be $1/G$. We have to prove that the numbers $\tfrac{5}{3}$, $\tfrac{8}{5}$, $\tfrac{11}{7}$, ..., after a certain ascertainable one, come closer to $\tfrac{3}{2}$ than $1/G$; that is, that the difference between $(3n+2)/(2n+1)$ and $\tfrac{3}{2}$ is less than $1/G$ for all values of n greater than some ascertainable number N.

Now

$$\frac{3n+2}{2n+1}-\frac{3}{2} = \frac{2(3n+2)-3(2n+1)}{2(2n+1)}$$

$$= \frac{1}{2(2n+1)},$$

and this difference is less than $1/G$ whenever

$$\frac{1}{2(2n+1)} < \frac{1}{G},$$

or

$$2(2n+1) > G,$$

or

$$4n > G-2,$$

or

$$n > \tfrac{1}{4}(G-2).$$

Thus, whatever value is selected for G, it is possible to make $(3n+2)/(2n+1)$ closer than $1/G$ to the limit $\tfrac{3}{2}$ by taking n greater than $\tfrac{1}{4}(G-2)$, and this is true for all such values of n. Hence

$$\lim_{n\to\infty} \frac{3n+2}{2n+1} = \frac{3}{2}.$$

Examples 1

1. Find a number N such that $1/10^n$ is less than $\frac{1}{100000}$ whenever $n \geqslant N$.

2. Find a number N such that $(\tfrac{1}{2})^n$ is less than $\frac{1}{4000}$ whenever $n \geqslant N$.

3. Find a number N such that $(\tfrac{1}{3})^n$ is less than $\frac{1}{1000}$ whenever $n \geqslant N$.

4. Find a number N such that $n/10^n$ is less than (i) $\frac{1}{100}$, (ii) $\frac{1}{1000}$, (iii) $\frac{1}{10000}$ whenever $n \geqslant N$.

5. Find a number N such that $n/2^n$ is less than (i) $\frac{1}{60}$, (ii) $\frac{1}{120}$, (iii) $\frac{1}{1000}$ whenever $n \geqslant N$.

6. Prove that

(i) $\lim_{n\to\infty} \frac{n+1}{n} = 1$, (ii) $\lim_{n\to\infty} \frac{n+1}{n^2} = 0$,

(iii) $\lim_{n\to\infty} \frac{4n+3}{2n+1} = 2$, (iv) $\lim_{n\to\infty} \frac{n^2+n+1}{n^2+1} = 1$.

2. The limit of k^n, nk^n, when $0 < k < 1$. Suppose that k is any positive number whose value is less than 1. We prove that (i) $k^n \to 0$, (ii) $nk^n \to 0$, *as* $n \to \infty$.

We take for gauge the number $1/G$, where G is regarded as large and positive. We have to show that there are ascertainable values of n such that (i) k^n, (ii) nk^n are less than $1/G$ for all greater values of n. To avoid trouble with fractions, we write

$$k = \frac{1}{1+p},$$

where $p > 0$. (This ensures that $k < 1$.) Then

$$k^n = \frac{1}{(1+p)^n}$$

$$= \frac{1}{1+{}_nC_1p+{}_nC_2p^2+\ldots+p^n}.$$

Now all the terms of the denominator are positive, so the value of the fraction is increased if any of these are omitted. In particular,

(i) $$k^n < \frac{1}{{}_nC_1p} = \frac{1}{np}.$$

If, then, we can choose n so large that

$$\frac{1}{np} < \frac{1}{G}$$

for all values of n greater than some ascertainable one, then we shall have the inequalities

$$k^n < \frac{1}{np} < \frac{1}{G}.$$

Now the condition on n is satisfied if

$$np > G,$$

or
$$n > \frac{G}{p}.$$
Hence
$$k^n < \frac{1}{G}$$
whenever $n > G/p$, so that, since G/p is an ascertainable number,
$$\lim_{n\to\infty} k^n = 0.$$

(ii) $k^n < \dfrac{1}{{}_nC_2\, p} = \dfrac{2}{n(n-1)p}$.

If, then, we can choose n so large that
$$\frac{2}{(n-1)p} < \frac{1}{G}$$
for all values of n greater than some ascertainable one, then we shall have the inequalities
$$nk^n < \frac{2}{(n-1)p} < \frac{1}{G}.$$

Now the condition on n is satisfied if
$$\tfrac{1}{2}(n-1)p > G,$$
or
$$n-1 > \frac{2G}{p},$$
or
$$n > 1+\frac{2G}{p}.$$
Hence
$$nk^n < \frac{1}{G},$$
whenever $n > 1+\dfrac{2G}{p}$, so that
$$\lim_{n\to\infty} nk^n = 0.$$

COROLLARIES. (i) If A is a *fixed* number, not depending on n, then, for $0 < k < 1$,
$$\lim_{n\to\infty} Ak^n = 0, \quad \lim_{n\to\infty} Ank^n = 0.$$

(ii) Since the *magnitude* of k^n is not affected by the sign of k, it is equally true that, *if k is any positive or negative number such that*
$$|k| < 1,$$
then
$$\lim_{n\to\infty} k^n = 0, \quad \lim_{n\to\infty} nk^n = 0.$$

3. The limit of $k^n/n!$* Suppose that k is any positive number. We prove that *$k^n/n! \to 0$ as $n \to \infty$.*

Suppose that N is the first integer greater than k, so that N is ascertainable. Then we may write $k^n/n!$ in the form

$$\frac{k}{1}\cdot\frac{k}{2}\cdot\ldots\cdot\frac{k}{N-1}\cdot\frac{k}{N}\cdot\frac{k}{N+1}\cdot\ldots\cdot\frac{k}{n},$$

where we have assumed that n has been chosen to exceed N. Now

$$\frac{k}{1}\cdot\frac{k}{2}\cdot\ldots\cdot\frac{k}{N-1}$$

is a definite number, independent of n, which we may call A. Also each of the numbers

$$\frac{k}{N+1}, \quad \ldots, \quad \frac{k}{n}$$

is less than k/N, which is itself less than 1.

Write

$$\frac{k}{N} = k' \quad (k' < 1),$$

and

$$n-N+1 = n'.$$

Then $n' \to \infty$ as $n \to \infty$. Also

$$\frac{k^n}{n!} < A\left(\frac{k}{N}\right)^{n-N+1}$$

$$< A(k')^{n'}.$$

But (p. 200)

$$A(k')^{n'} \to 0,$$

and so

$$\frac{k^n}{n!} \to 0.$$

COROLLARY. Since sign does not affect magnitude, it follows that, *if k is any number whatever, then*

$$\frac{k^n}{n!} \to 0 \quad as \quad n \to \infty.$$

4. The geometric series. We have proved (p. 185) that the sum S_n of the first n terms of the geometric series

$$a+ar+ar^2+\ldots$$

is

$$\frac{a(1-r^n)}{1-r}.$$

* To be omitted at first reading.

Thus
$$S_n = \frac{a}{1-r} - \frac{ar^n}{1-r},$$

so that
$$\frac{a}{1-r} - S_n = \left(\frac{a}{1-r}\right)r^n.$$

Suppose now that the value of r is numerically less than 1, so that
$$|r| < 1.$$

Then, by §2,* if $1/G$ is any gauge number, a value of n can be found such that
$$\left|\left(\frac{a}{1-r}\right)r^n\right| < \frac{1}{G}$$

for all larger values of n. Hence, for such values of n,
$$\left|\frac{a}{1-r} - S_n\right| < \frac{1}{G}.$$

Thus the sequence $S_1, S_2, S_3, \ldots$ is such that all its members, after a certain ascertainable one, come closer to the number $a/(1-r)$ than any gauge number $1/G$ we care to select. Hence
$$S_n \to \frac{a}{1-r} \quad \text{as} \quad n \to \infty.$$

We call
$$\frac{a}{1-r}$$

the *sum to infinity* of the geometric series. *For its existence, it is* essential *that*
$$|r| < 1.$$

It is conventional to write
$$a + ar + ar^2 + \ldots + ar^{n-1}$$

to denote a terminating series of n terms, and
$$a + ar + ar^2 + \ldots$$

to denote an 'infinite' series. Sometimes general terms are inserted for clarity:
$$a + ar + ar^2 + \ldots + ar^{p-1} + \ldots + ar^{n-1}$$

and
$$a + ar + ar^2 + \ldots + ar^{p-1} + \ldots.$$

Note that, although we talk about *infinite series* and *sums to infinity*, we are not, in any sense, adding an 'infinite' number of terms; that is a meaningless operation. What we do is to sum the series for a finite

* The collorary, with $A \equiv a/(1-r)$.

number n of terms, obtaining a sum S_n, and then to examine whether the expression S_n tends to any limit S (in the sense defined on p. 198) as n tends to infinity.

Illustration 2. *To exhibit graphically the sum of the geometric series*

$$1+\tfrac{1}{2}+\tfrac{1}{4}+\tfrac{1}{8}+\tfrac{1}{16}+\ldots.$$

The sum S_n of the first n terms is

$$\frac{1-(\frac{1}{2})^n}{1-\frac{1}{2}},$$

or

$$2-(\tfrac{1}{2})^{n-1},$$

so that the sum to infinity is 2.

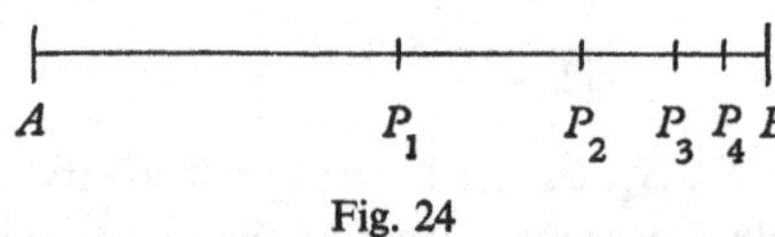

Fig. 24

Draw a straight line AP_1B in which the lengths of AP_1, P_1B in any convenient units are each 1 (fig. 24). Let P_2 be the middle point of P_1B; P_3 be the middle point of P_2B; P_4 be the middle point of P_3B; and so on. Then

$$AP_1 = 1,\quad P_1P_2 = \tfrac{1}{2},\quad P_2P_3 = \tfrac{1}{4},\quad P_3P_4 = \tfrac{1}{8},\quad P_4P_5 = \tfrac{1}{16},$$

and so on. Thus

$$S_1 = AP_1,\quad S_2 = AP_2,\quad S_3 = AP_3,\quad S_4 = AP_4,\quad \ldots,\quad S_n = AP_n.$$

The point P_n gets closer and closer to B as n increases, but never actually gets there; on the other hand, the distance P_nB can be made less than any number we care to assign. Thus the limit of S_n is the length AB, so that

$$S = AB.$$

Illustration 3. *To find the sum of the first n terms, and the sum to infinity, of the series whose p-th term is* px^{p-1}, *where* $-1 < x < 1$.

If S_n is the sum of the first n terms, then

$$S_n = 1+2x+3x^2+\ldots+nx^{n-1}.$$

We sum this series by a device similar to that used for the geometric series itself:

$$xS_n = \qquad x+2x^2+\ldots+(n-1)x^{n-1}+nx^n.$$

Subtract:

$$(1-x)S_n = 1+x+x^2+\ldots+x^{n-1}-nx^n$$

$$= \frac{1-x^n}{1-x}-nx^n$$

on summing the geometric series. Hence

$$S_n = \frac{1-x^n}{(1-x)^2} - \frac{nx^n}{(1-x)}$$

$$= \frac{1}{(1-x)^2} - \frac{x^n}{(1-x)^2} - \frac{nx^n}{(1-x)}.$$

Now, by the Corollaries in §2 (p. 201),

$$Ax^n \to 0, \quad Bnx^n \to 0 \quad (-1 < x < 1),$$

where, for the present argument,

$$A = 1/(1-x)^2, \quad B = 1/(1-x),$$

with $x \neq 1$. Hence
$$S_n \to \frac{1}{(1-x)^2},$$

so that *the sum of the infinite series*

$$1+2x+3x^2+\ldots$$

is $1/(1-x)^2$.

GENERALIZATION. It may be worth while to consider briefly the series whose sum is

$$S_n \equiv 1.2+2.3x+3.4x^2+\ldots+n(n+1)x^{n-1}.$$

Then

$$xS_n \equiv \qquad 1.2x+2.3x^2+\ldots+(n-1)nx^{n-1}+n(n+1)x^n.$$

Subtract:

$$(1-x)S_n = 1.2+2.2x+3.2x^2+\ldots+n.2x^{n-1}-n(n+1)x^n$$

$$= 2(1+2x+3x^2+\ldots+nx^{n-1}) - n(n+1)x^n$$

$$= \frac{2}{(1-x)^2} - \frac{2x^n}{(1-x)^2} - \frac{2nx^n}{(1-x)} - n(n+1)x^n.$$

Thus
$$S_n = \frac{2}{(1-x)^3} - \frac{2x^n}{(1-x)^3} - \frac{2nx^n}{(1-x)^2} - \frac{n(n+1)x^n}{(1-x)}.$$

We can show, by reasoning similar to that used before, that, if x lies between -1 and $+1$, the sum to infinity S exists, and that

$$S = \frac{2}{(1-x)^3}.$$

Thus
$$1+\frac{2.3}{2}x+\frac{3.4}{2}x^2+\frac{4.5}{2}x^3+\ldots = (1-x)^{-3}.$$

Examples 2

1. Prove that

$$\frac{1}{1.2}+\frac{1}{2.3}+\frac{1}{3.4}+\ldots+\frac{1}{n(n+1)} = 1-\frac{1}{n+1},$$

and deduce that the sum of the infinite series

$$\frac{1}{1.2}+\frac{1}{2.3}+\frac{1}{3.4}+\ldots$$

is 1.

2. Prove that, if x lies between -1 and 1, then the series

$$1+2x+3x^2+4x^3+\ldots$$

has sum to infinity $1/(1-x)^2$.

3. Prove that the sum of the series

$$\frac{1}{2.3.4.5}+\frac{4}{3.4.5.6}+\ldots+\frac{n^2}{(n+1)(n+2)(n+3)(n+4)}+\ldots$$

is 5/36.

4. Prove that the sum to n terms of the series

$$\frac{1}{2.3.4}+\frac{1}{3.4.5}+\frac{1}{4.5.6}+\ldots$$

is

$$\frac{1}{12}-\frac{1}{2(n+2)(n+3)},$$

and hence find the sum to infinity.

5. Find the sum of n terms of the series whose rth term is

$$\frac{2r-1}{r(r+1)(r+2)}.$$

Find whether the sum to infinity exists.

6. Prove that

$$\frac{1}{1.2.3}+\frac{1}{2.3.4}+\ldots+\frac{1}{(n-1)n(n+1)} = \frac{(n-1)(n+2)}{4n(n+1)}.$$

Show that the series has a sum to infinity and find roughly how many terms must be taken to give a sum differing from the sum to infinity by not more than one part in a million.

7. Show that the sum of the first n terms of the series

$$\frac{2}{1.3}+\frac{2}{3.5}+\frac{2}{5.7}+\dots$$

is $1-1/(2n+1)$.

Deduce that the infinite series converges, and find its sum.

17

THE BINOMIAL SERIES

1. Statement. We have proved (p. 166) that, when n is a positive integer,

$$(1+x)^n = 1+nx+\frac{n(n-1)}{2!}x^2+\frac{n(n-1)(n-2)}{3!}x^3+\dots,$$

the series consisting of $n+1$ terms. It is of interest to consider what meaning can be given to the series

$$1+nx+\frac{n(n-1)}{2!}x^2+\dots+\frac{n(n-1)\dots(n-r+1)}{r!}x^r+\dots$$

even when n is not a positive integer.

The series is an infinite series, for it terminates only if one of the factors

$$n, n-1, n-2, \dots, n-r+1, \dots$$

becomes zero for some value of r. We cannot therefore be sure that it has any meaning at all. For example, when $n = -1$ the series is

$$1-x+x^2-x^3+x^4-\dots,$$

which, with $x = \frac{1}{2}$, is a geometric series having sum to infinity $\frac{2}{3}$, but which, with $x = -2$, is the meaningless series

$$1+2+4+8+16+\dots.$$

On the other hand, the case $n = -1$ is suggestive, for, when

$$-1 < x < 1,$$

the result is the infinite series

$$1-x+x^2-x^3+\dots$$

whose sum (p. 202) is the known function

$$\frac{1}{1-(-x)},$$

or

$$(1+x)^{-1}.$$

Again, when $n = -2$ the series is

$$1-2x+\frac{(-2)(-3)}{2!}x^2+\frac{(-2)(-3)(-4)}{3!}x^3+\ldots = 1-2x+3x^2-4x^3+\ldots,$$

which (p. 204, Illustration 3), for x between -1 and 1, has sum

$$(1+x)^{-2}.$$

In fact, generalization by induction of the method given on p. 205 leads to the result that, *when n is a negative integer, the sum of the infinite series*

$$1+nx+\frac{n(n-1)}{2!}x^2+\frac{n(n-1)(n-2)}{3!}x^3+\ldots$$

for $-1 < x < 1$ *is* $$(1+x)^n.$$

More generally, it would seem that the infinite series

$$1+nx+\frac{n(n-1)}{2!}x^2+\frac{n(n-1)(n-2)}{3!}x^3+\ldots$$

may have a sum whose value, when existent, is equal to

$$(1+x)^n.$$

This is perhaps proved most easily by the methods of the calculus. See, for example, the author's *An Analytical Calculus*, vol. II, p. 52. We therefore assume without further proof here the BINOMIAL SERIES

$$(1+x)^n = 1+nx+\frac{n(n-1)}{2!}x^2+\ldots+\frac{n(n-1)\ldots(n-r+1)}{r!}x^r+\ldots,$$

valid for all values of x in the interval $-1 < x < 1$.

For example,

$$\sqrt{(1+x)} = (1+x)^{\frac{1}{2}} = 1+\tfrac{1}{2}x+\frac{\frac{1}{2}(-\frac{1}{2})}{2!}x^2+\frac{\frac{1}{2}(-\frac{1}{2})(-\frac{3}{2})}{3!}x^3+\ldots$$

$$= 1+\tfrac{1}{2}x-\frac{1}{2.4}x^2+\frac{1.3}{2.4.6}x^3-\frac{1.3.5}{2.4.6.8}x^4+\ldots.$$

When n is fractional, there is a possible ambiguity in the interpretation of $(1+x)^n$; thus $(1+x)^{\frac{1}{2}}$ has two square roots. When such ambiguity exists, *that interpretation must be selected which gives the value* $+1$ *when* $x = 0$, this being the corresponding value of the series.

It is most important to observe that the binomial series is defined for the expansion of the expression $1+x$, where the first term is 1. In other cases the procedure of *Illustration* 1, p. 210, must be adopted.

Illustration 1. *To find an expansion for*

$$(5-4x)^{\frac{1}{3}}.$$

The expression is
$$5^{\frac{1}{3}}(1-\tfrac{4}{5}x)^{\frac{1}{3}},$$
and this, on expansion, is

$$5^{\frac{1}{3}}\left\{1+\tfrac{1}{3}(-\tfrac{4}{5}x)+\frac{\frac{1}{3}(-\frac{2}{3})}{2!}(-\tfrac{4}{5}x)^2+\frac{\frac{1}{3}(-\frac{2}{3})(-\frac{5}{3})}{3!}(-\tfrac{4}{5}x)^3+\dots\right\}$$

$$=5^{\frac{1}{3}}\left\{1-\frac{4}{3.5}x-\frac{2.4^2}{6.3.5^2}x^2-\frac{2.5.4^3}{9.6.3.5^3}x^3-\dots\right\}.$$

2. More elaborate combinations of expansions. It is sometimes necessary to find the first few terms in an expansion even though the more general treatment presents difficulty. This usually involves *multiplication of series*, which is a hard idea for rigorous treatment (cf. p. 172 for a simpler case). The actual calculation of the answer is, however, comparatively simple, and that must suffice for the present.

For example, it may be required *to find the first four terms in the expansion of*

$$E\equiv\frac{(1+2x)^{\frac{1}{2}}}{(1+3x^2)^{\frac{1}{3}}}$$

in a series of ascending powers of x. Then

$$(1+2x)^{\frac{1}{2}}=1+\tfrac{1}{2}(2x)+\frac{\frac{1}{2}(-\frac{1}{2})}{2}(2x)^2+\frac{\frac{1}{2}(-\frac{1}{2})(-\frac{3}{2})}{6}(2x)^3+\dots$$

$$=1+x-\tfrac{1}{2}x^2+\tfrac{1}{2}x^3+\dots,$$

and
$$(1+3x^2)^{-\frac{1}{3}}=1+(-\tfrac{1}{3})(3x^2)+\frac{(-\frac{1}{3})(-\frac{4}{3})}{2}(3x^2)^2+\dots$$
$$=1-x^2+2x^4+\dots,$$

so that, *for powers of x up to x^3*,

$$E=(1+x-\tfrac{1}{2}x^2+\tfrac{1}{2}x^3)(1-x^2).$$

Hence
$$E=1+x-\tfrac{1}{2}x^2+\tfrac{1}{2}x^3-x^2-x^3$$
$$=1+x-\tfrac{3}{2}x^2-\tfrac{1}{2}x^3,$$
to that order.

Revision Examples XI

1. Prove that, if x is so small that its fourth power may be neglected,

$$\frac{(1-\frac{1}{2}x)^2(1+2x)^{\frac{1}{2}}}{(1+3x^2)^{\frac{1}{3}}}=1-\tfrac{9}{4}x^2+\tfrac{5}{4}x^3.$$

2. Show that the expansion of $(1-x)^{\frac{1}{3}}(1+2x)^{-\frac{3}{4}}$ begins with the terms $1-\frac{11}{6}x+\frac{217}{72}x^2$, and find the coefficient of x^3.

3. Expand $(1-\frac{1}{4}x^4)^{-4}$ in ascending powers of x as far as the term containing x^{12}.

4. Use the binomial theorem to expand both of the expressions

$$\sqrt{(1-2x)}, \quad \frac{2-3x}{2-x}$$

in ascending powers of x as far as the term in x^4 in each case. State the range of values of x for which each of the expansions is valid.

By putting $x = \frac{1}{50}$ in your expansions, prove that $\sqrt{6}$ differs from $\frac{485}{198}$ by less than 0·00001.

5. Use the binomial theorem to calculate the value of $(27{\cdot}27)^{-\frac{1}{3}}$ to 6 places of decimals.

6. Expand $$(1-2x)^{-\frac{1}{2}}$$ in ascending powers of x as far as the term in x^3.

7. Use the binomial theorem to calculate the value of $(16{\cdot}32)^{\frac{1}{4}}$ to 6 places of decimals.

8. Find the coefficient of x^r in the expansion of

$$\frac{1+3x}{(1-x)^3} \quad (-1 < x < 1).$$

9. Use the binomial series to evaluate $(1{\cdot}1)^{-3}$ to 3 significant figures.

10. Determine the value of k if the expansion of

$$\frac{1-kx^2}{(1-x^2)^{\frac{1}{2}}}$$

in powers of x has no term in x^2; evaluate the three following terms for this particular case.

11. Write down the first five terms in the expansion of $(1+x)^{-\frac{1}{2}}$. The time of swing of a pendulum of length l is given by the formula

$$T = \pi\sqrt{(l/g)}$$

seconds. Show that, if the pendulum in a clock which normally beats seconds is increased in length by one-tenth per cent., the clock will lose approximately $43\frac{1}{4}$ seconds per day.

12. Determine a, b, c so that the expansion of

$$\frac{a+bx+cx^2}{(1-x)^4}$$

begins with the terms $1^3+2^3x+3^3x^2$, and show that the coefficient of x^n is then $(n+1)^3$.

13. Use the binomial theorem to evaluate $(1\cdot 06)^{\frac{7}{5}}$ to six significant figures.

14. Evaluate $\sqrt{(99)}$ correct to six places of decimals.

15. Prove that the coefficient of x^{2n} in the expansion of

$$(1+x^2)^n(1-x)^{-3}$$

in ascending powers of x is

$$(n^2+4n+2)2^{n-1}.$$

16. Prove that the coefficient of x^{2n} in the expansion of

$$(1+x^2)^n(1-x)^{-4}$$

in ascending powers of x is

$$\tfrac{1}{3}(n+2)(n^2+7n+3)2^{n-1}.$$

17. Find the condition that the nth term in the expansion of

$$(1-x)^{-k}$$

exceeds the next, assuming that $k > 0$ and $0 < x < 1$.

Hence find which are the greatest terms in the expansions of

$$\text{(i) } (1-\tfrac{5}{6})^{-5}, \qquad \text{(ii) } (1-\tfrac{1}{10})^{-9}.$$

18. Prove that the sum to infinity

$$1+\frac{1}{3}+\frac{1.3}{2!\,3^2}+\frac{1.3.5}{3!\,3^3}+\ldots$$

is $\sqrt{3}$.

19. If the expression

$$\frac{1}{1-x}-\frac{3}{(1-2x)^{\frac{1}{2}}}+\frac{3}{(1-3x)^{\frac{1}{3}}}-\frac{1}{(1-4x)^{\frac{1}{4}}}$$

is expanded in the form

$$a_0+a_1x+a_2x^2+\ldots,$$

find the coefficients a_0, a_1, a_2, a_3, a_4.

20. Prove that

$$\left(\frac{1+x}{1-x}\right)^3 = 1+6x+18x^2+\ldots+(4n^2+2)x^n+\ldots.$$

3. Expansion of rational functions. The binomial series can be used to expand rational functions in series of ascending powers of x; the range of x for convergence must then be examined very carefully.

For example, let

$$E \equiv \frac{x^2+20}{(x-2)^2(x+4)}.$$

The expression for E in partial fractions is

$$E \equiv \frac{4}{(x-2)^2}+\frac{1}{x+4}.$$

In preparation for binomial expansions (compare the remark on p. 209), write

$$E \equiv \frac{1}{(1-\frac{1}{2}x)^2}+\frac{1}{4(1+\frac{1}{4}x)}$$

$$\equiv (1-\tfrac{1}{2}x)^{-2}+\tfrac{1}{4}(1+\tfrac{1}{4}x)^{-1}.$$

Then, expanding each binomial expression,

$$E \equiv 1+2(\tfrac{1}{2}x)+3(\tfrac{1}{2}x)^2+\ldots+(n+1)(\tfrac{1}{2}x)^n+\ldots$$
$$+\tfrac{1}{4}-\tfrac{1}{4}(\tfrac{1}{4}x)+\tfrac{1}{4}(\tfrac{1}{4}x)^2+\ldots+\tfrac{1}{4}(-1)^n(\tfrac{1}{4}x)^n+\ldots,$$

so that the coefficient of x^n in the expansion is

$$\frac{(n+1)}{2^n}+\frac{(-1)^n}{4^{n+1}}.$$

Moreover, the expansions are valid only for the common part of the intervals

$$-1 < \tfrac{1}{2}x < 1, \quad -1 < \tfrac{1}{4}x < 1,$$

that is,

$$-2 < x < 2, \quad -4 < x < 4.$$

To satisfy both, the range is

$$-2 < x < 2.$$

4. Application of the binomial theorem to partial fractions with high exponent. The methods given in chapter 10 for calculating partial fractions become somewhat lengthy when the powers are high. An alternative method may then be used based on the binomial theorem for negative integers.

Consider, as an example, the function

$$E \equiv \frac{1+x^3}{x^6(1+2x^2)}.$$

We require *the six partial fractions corresponding to x*. The argument may, in illustration, be split into the two parts: calculation, justification.

(*a*) *Calculation.* Express the 'residual' function

$$\frac{1+x^3}{1+2x^2}$$

as a series of ascending powers of x, taking terms as far as $x^{6-1} = x^5$. This is

$$\begin{aligned}(1+x^3)(1+2x^2)^{-1} &= (1+x^3)(1-2x^2+4x^4-8x^6+\ldots)\\ &= 1-2x^2+x^3+4x^4-2x^5-8x^6+\ldots,\end{aligned}$$

for such values of x as make the series convergent (actually $|2x^2| < 1$, so that $|x| < 1/\sqrt{2}$). Hence

$$\frac{1+x^3}{x^6(1+2x^2)} = \frac{1}{x^6}-\frac{2}{x^4}+\frac{1}{x^3}+\frac{4}{x^2}-\frac{2}{x}+(\text{terms in } x^0, x^1, x^2, \ldots).$$

The suggestion is that the partial fractions in x are

$$\frac{1}{x^6}-\frac{2}{x^4}+\frac{1}{x^3}+\frac{4}{x^2}-\frac{2}{x}.$$

(*b*) *Justification.* (The process of justification in this simple case also succeeds in obtaining incidentally the complete resolution of the function into partial fractions.)

Form the difference

$$\begin{aligned}\frac{1+x^3}{x^6(1+2x^2)}-\frac{1}{x^6}+\frac{2}{x^4}-\frac{1}{x^3}-\frac{4}{x^2}+\frac{2}{x}\\ &= \frac{(1+x^3)-(1+2x^2)(1-2x^2+x^3+4x^4-2x^5)}{x^6(1+2x^2)}\\ &\equiv \frac{-8x^6+4x^7}{x^6(1+2x^2)} \equiv \frac{-8+4x}{1+2x^2}.\end{aligned}$$

Thus
$$\frac{1+x^3}{x^6(1+2x^2)} \equiv \frac{1}{x^6}-\frac{2}{x^4}+\frac{1}{x^3}+\frac{4}{x^2}-\frac{2}{x}-\frac{8-4x}{1+2x^2},$$

so that the suggestion was correct, and the resolution into partial fractions is completed.

More generally, consider the function

$$\frac{f(x)}{x^n g(x)},$$

where $g(x)$ does not contain x as a factor.

The expression in partial fractions assumes the form (chapter 10)

$$\frac{f(x)}{x^n g(x)} \equiv \frac{A_0}{x^n}+\frac{A_1}{x^{n-1}}+\dots+\frac{A_{n-1}}{x}+\frac{h(x)}{g(x)},$$

where $h(x)/g(x)$ is the part arising from the factors of $g(x)$. Multiply by x^n; then

$$\frac{f(x)}{g(x)} \equiv A_0+A_1x+\dots+A_{n-1}x^{n-1}+\frac{x^n h(x)}{g(x)}.$$

Now $h(x)/g(x)$ can, subject to convergence, be expanded (for example, by expanding each of the partial fractions into which it can be decomposed) as a series of ascending powers of x in the form

$$\frac{h(x)}{g(x)} \equiv b_0+b_1x+b_2x^2+\dots.$$

Hence, substituting this in the earlier expansion, we have

$$\frac{f(x)}{g(x)} \equiv A_0+A_1x+\dots+A_{n-1}x^{n-1}+b_0x^n+b_1x^{n+1}+\dots,$$

so that $A_0, A_1, \dots, A_{n-1}$ *are seen as the first* n *coefficients in the expansion of* $f(x)/g(x)$.

To find the partial fractions corresponding to $x+h$, where $h \neq 0$, a simple modification can be used. The substitution $y = x+h$ transforms the given rational function in x into one in y of precisely the type considered, and the calculations can be made in the new variable, transforming back to x at the end for the final answer.

***Illustration* 2.** *To express*

$$E \equiv \frac{1+10x^2+5x^4}{(1-x)^4(1+x)^3}$$

in partial fractions.

(i) *Corresponding to* $1-x$.

Write $$1-x = y,$$

so that $$x = 1-y.$$

Then $$E \equiv \frac{1+10(1-2y+y^2)+5(1-4y+6y^2-4y^3+y^4)}{y^4(2-y)^3}$$

$$\equiv \frac{16-40y+40y^2-20y^3+5y^4}{8y^4(1-\frac{1}{2}y)^3}.$$

As far as terms in y^3,

$$\begin{aligned} y^4E &\equiv \tfrac{1}{8}(16-40y+40y^2-20y^3+\dots)(1-\tfrac{1}{2}y)^{-3} \\ &\equiv \tfrac{1}{8}(16-40y+40y^2-20y^3+\dots)(1+\tfrac{3}{2}y+\tfrac{6}{4}y^2+\tfrac{10}{8}y^3+\dots) \\ &\equiv \tfrac{1}{8}(16-16y+4y^2+0.y^3+\dots) \\ &\equiv 2-2y+\tfrac{1}{2}y^2+0.y^3+\dots. \end{aligned}$$

Hence
$$E \equiv \frac{2}{y^4}-\frac{2}{y^3}+\frac{1}{2y^2}+\frac{0}{y}+\dots,$$

so that the corresponding partial fractions are

$$\frac{2}{(1-x)^4}-\frac{2}{(1-x)^3}+\frac{1}{2(1-x)^2}.$$

(ii) *Corresponding to* $1+x$.

Write
$$1+x = y,$$
so that
$$x = y-1.$$

Then
$$\begin{aligned} E &\equiv \frac{1+10(y^2-2y+1)+5(y^4-4y^3+6y^2-4y+1)}{(2-y)^4y^3} \\ &\equiv \frac{16-40y+40y^2-20y^3+5y^4}{16y^3(1-\frac{1}{2}y)^4}. \end{aligned}$$

As far as terms in y^2,

$$\begin{aligned} y^3E &\equiv \tfrac{1}{16}(16-40y+40y^2+\dots)(1-\tfrac{1}{2}y)^{-4} \\ &\equiv \tfrac{1}{2}(2-5y+5y^2+\dots)(1+\tfrac{4}{2}y+\tfrac{10}{4}y^2+\dots) \\ &\equiv \tfrac{1}{2}(2-y+0.y^2+\dots) \\ &\equiv 1-\tfrac{1}{2}y+0.y^2+\dots. \end{aligned}$$

Hence
$$E \equiv \frac{1}{y^3}-\frac{1}{2y^2}+\frac{0}{y}+\dots,$$

so that the corresponding partial fractions are

$$\frac{1}{(1+x)^3}-\frac{1}{2(1+x)^2}.$$

In all
$$E \equiv \frac{2}{(1-x)^4}-\frac{2}{(1-x)^3}+\frac{1}{2(1-x)^2}+\frac{1}{(1+x)^3}-\frac{1}{2(1+x)^2}.$$

***Illustration* 3.** A similar method may be used when one or more factors in the denominator of a rational function occurs to a power designated by a general symbol such as n. For example, *to express the function*

$$E \equiv \frac{1}{(1-x)^n(2-x)^2}$$

in partial fractions.

(i) *Corresponding to* $2-x$.

Write
$$2-x=y,$$
so that
$$x=2-y.$$
Then
$$E\equiv\frac{1}{(y-1)^n y^2},$$
so that
$$y^2E\equiv(y-1)^{-n}$$
$$\equiv(-1)^n(1-y)^{-n}.$$
As far as the term in y,
$$y^2E\equiv(-1)^n(1+ny+\ldots),$$
so that
$$E\equiv\frac{(-1)^n}{y^2}+\frac{(-1)^n n}{y}+\ldots,$$
and the corresponding partial fractions are
$$\frac{(-1)^n}{(2-x)^2}+\frac{(-1)^n n}{2-x}.$$

(ii) *Corresponding to* $1-x$.

Write
$$1-x=y,$$
so that
$$x=1-y.$$
Then
$$E\equiv\frac{1}{y^n(1+y)^2},$$
so that
$$y^nE\equiv(1+y)^{-2}$$
$$\equiv 1-2y+3y^2-4y^3+\ldots.$$
Hence
$$E\equiv\frac{1}{y^n}-\frac{2}{y^{n-1}}+\frac{3}{y^{n-2}}-\ldots+\frac{(-1)^{r+1}r}{y^{n-r+1}}+\ldots+\frac{(-1)^{n+1}n}{y}+\ldots,$$
so that the corresponding partial fractions are
$$\frac{1}{(1-x)^n}-\frac{2}{(1-x)^{n-1}}+\ldots+\frac{(-1)^{r+1}r}{(1-x)^{n-r+1}}+\ldots+\frac{(-1)^{n+1}n}{1-x}.$$

In all,
$$E\equiv\frac{(-1)^n}{(2-x)^2}+\frac{(-1)^n n}{2-x}+\frac{1}{(1-x)^n}-\frac{2}{(1-x)^{n-1}}+\ldots+\frac{(-1)^{n+1}n}{1-x}.$$

Revision Examples XII

1. Express
$$\frac{3x^4+12x+8}{(x+1)^5}$$
in partial fractions.

2. Express
$$\frac{x}{(x-2)^4(x+1)(x-1)}$$
in partial fractions.

3. Prove that the coefficient of x^{4n} in the expansion of

$$\frac{1}{(1-x)(1-x^2)(1-x^4)}$$

in ascending powers of x is $(n+1)^2$.

4. Express
$$\frac{2x+1}{(x-2)(x+1)^2}$$
in the form
$$\frac{A}{x-2}+\frac{B}{(x+1)^2}+\frac{C}{x+1}.$$

Deduce that, when the given expression is expanded in ascending powers of x, the coefficient of x^n is

$$-\frac{1}{9}\left\{\frac{5}{2^{n+1}}\pm(3n-2)\right\},$$

the sign + or − being taken according as n is odd or even.

5. Determine a, b, c so that

$$\frac{1+2x}{(1+x^2)(1-3x)}\equiv\frac{a}{1-3x}+\frac{bx+c}{1+x^2},$$

and show that, if the function is expanded in ascending powers of x, the coefficient of x^n is

$$\begin{cases}\frac{1}{2}(3^{n+1}+1) & \text{if } n \text{ is of the form } 4r+1 \text{ or } 4r+2,\\ \frac{1}{2}(3^{n+1}-1) & \text{if } n \text{ is of the form } 4r \text{ or } 4r+3.\end{cases}$$

6. Express the function $\dfrac{x^2+1}{(x+2)(x-1)^2}$

in partial fractions, and hence find the first four terms of the expansion of the function as a series of ascending powers of x, stating the necessary restrictions as to the values of x.

7. Express
$$\frac{4x^2-x}{(x-1)(x^2+2)}$$
in partial fractions. Hence obtain the first four terms in the expansion of the function in ascending powers of x, stating the necessary restrictions on the values of x.

8. Express
$$\frac{1}{(1+2x)(1+x^2)}$$
in partial fractions, and obtain the coefficients of (i) x^6, (ii) x^9 in the expansion of the function in a series of ascending powers of x.

9. Express in partial fractions

$$\frac{x}{(x-1)(x-2)(x-3)}, \quad \frac{x}{(x-2)^3}.$$

Find an expression for the coefficient of x^n in the expansion of

$$\frac{1}{(x-1)(x-2)(x-3)}$$

as a series of ascending powers of x. For what values of x does the series converge?

10. Resolve the expression

$$\frac{9}{(x-1)(x+2)^2}$$

into partial fractions and find the general term of this expression when expanded in ascending powers of x.

11. Express $$\frac{20}{(x^2+x+4)(x-2)^2}$$

in partial fractions, and expand this expression in ascending powers of x as far as the term in x^4.

12. Prove that, if x has not one of the values $-1, -2, \ldots, -n$,

$$\frac{(x-1)(x-2)\ldots(x-n)}{(x+1)(x+2)\ldots(x+n)} \equiv 1 + \sum_{r=1}^{n} \frac{(-1)^{n-r+1}(n+r)!}{(n-r)!\,r!(r-1)!\,(x+r)}.$$

13. Prove that

$$\sum_{r=1}^{n} \frac{2r}{(2r-1)(2r+1)(2r+3)} = \frac{n(n+1)}{(2n+1)(2n+3)}.$$

14. Given that

$$\frac{ax^2+bx+c}{(x-\alpha)(x-\beta)(x-\gamma)} \equiv \frac{A}{x-\alpha} + \frac{B}{x-\beta} + \frac{C}{x-\gamma},$$

find the condition that $A+B+C = 0$.

Evaluate $$\sum_{n=1}^{N} \frac{3n-1}{n(n+1)(n+3)}.$$

15. Express $$\frac{6x^2+1}{(x-1)^2(x-3)(x^2+2)}$$

in partial fractions, and expand as a series of increasing powers of x, stating the coefficients of x^{2n} and x^{2n+1}.

16. Express
$$\frac{x^3+x^2-x+1}{(x+1)^4}$$
in partial fractions.

17. Express in partial fractions
$$\frac{4x^3-12x^2+14x-7}{(x^2-3x+2)^3},$$
and show that the coefficient of x^n in the expansion of this expression in ascending powers of x is
$$-\frac{n(n+1)}{2}-\frac{(n+1)(5n+14)}{2^{n+4}}.$$
State the range of values of x for which this expression is valid.

18. Find the partial fractions of
$$\frac{x}{(x+2)(x-1)^n},\quad \frac{x}{(x-1)(2x-1)(x-2)^n},$$
where n is a positive integer.

19. Express
$$\frac{1}{x(x-2)(x-1)^n}$$
as the sum of elementary partial fractions, where n is a positive integer; distinguish between the cases n even and n odd.

20. Express
$$\frac{4x^2}{(x^2+1)(x-1)^3}$$
in partial fractions, and prove that, if $|x| < 1$, the coefficient of x^{2n} in the expansion of the expression in ascending powers of x is
$$1-2n-4n^2+(-1)^{n-1}.$$

21. Express in partial fractions
$$\frac{(2a)^n}{(x^2-a^2)^n},$$
where n is a positive integer.

18

THE EXPONENTIAL SERIES

[Beginners will probably find some of this work hard, but, if time allows, the effort to grasp it is worth while. The summarized statement in §4 may be found sufficient at first.]

1. Genesis. It is not easy to make clear, at the present level, just why the two series to be discussed in this and the following chapter play such a fundamental role in mathematics. This brief sketch is intended to indicate what is involved, without undue emphasis on rigour. Alternative treatments will probably be studied in the calculus course.

The two ideas of *exponent* and *logarithm* are closely related in elementary mathematics. In an expression like

$$a^m,$$

the number m is called the *exponent* of the power to which a is raised; and, if

$$a^m = p,$$

then m is called the *logarithm* of p to the base a, written

$$m = \log_a p.$$

It is understood throughout this section that *the value of a is positive.* The rule

$$a^m \times a^n = a^{m+n}$$

states that *powers of a are multiplied by adding exponents*; its expression in terms of logarithms is that, if

$$p = a^m, \quad q = a^n,$$

so that

$$pq = a^{m+n},$$

then *numbers are multiplied by adding logarithms*:

$$\log_a p + \log_a q = \log_a pq.$$

In numerical work, logarithms are usually evaluated to the base 10. That, however, is a somewhat fortuitous number depending (perhaps) on the number of fingers on the human hands. The work which follows is the result of an attempt to find a more mathematically natural base number—though not everyone will feel that the *dénouement*

$$1+1+\frac{1}{2!}+\frac{1}{3!}+\frac{1}{4!}+\dots$$

is much more natural than 10.

2. The exponential series. We begin by a consideration of the expression

$$a^m$$

for given positive a and varying values of m. It is a function of m, which may be denoted temporarily by the more general notation

$$f(m).$$

The rule $a^m \times a^n = a^{m+n}$ is expressed in the new notation by the identical relation

$$f(m)f(n) \equiv f(m+n)$$

for all pairs of values of m, n.

In particular, if $m = 0$, then

$$f(0)f(n) \equiv f(n),$$

so that *either* $f(n) \equiv 0$, a possibility which can be ignored as trivial, *or* $f(0) = 1$.

The identity

$$f(m)f(n) \equiv f(m+n)$$

holds for all pairs of values of m, n. Make now the assumption that the function $f(x)$ is of a type that can be differentiated. If the number m is kept constant and the identity differentiated with respect to the variable n, then

$$\begin{aligned} f(m)f'(n) &\equiv f'(m+n)\frac{d(m+n)}{dn} \\ &= f'(m+n). \end{aligned}$$

In particular, the value $n = 0$ gives the identity

$$f'(m) \equiv f'(0)f(m).$$

The value of $f'(0)$ is a numerical constant, and it is independent of m; denote its value by k. Then

$$f'(m) = kf(m).$$

Continued differentiation yields the relation

$$f''(m) = kf'(m) = k^2 f(m),$$
$$f'''(m) = k^2 f'(m) = k^3 f(m),$$
$$f^{(\text{iv})}(m) = k^3 f'(m) = k^4 f(m),$$

and so on. Thus (so long as the function is of a type which can be differentiated repeatedly)

$$f^{(p)}(m) = k^p f(m).$$

In particular, giving m the value zero, we have the relation

$$f^{(p)}(0) = k^p f(0)$$
$$= k^p.$$

[At this stage the mathematical equipment of the reader seems likely to be inadequate, but the assumptions to be made will, we hope, appear reasonable. One of the best ways of examining a function is to express it in terms of an infinite series.]

We now consider whether it is possible to find an expression for the function $f(m)$ in the form of an infinite series of ascending powers of m. If so, the expression would be

$$f(m) \equiv 1 + a_1 m + a_2 m^2 + a_3 m^3 + \ldots + a_p m^p + \ldots,$$

the first term being 1 since $f(0) = 1$.

In order to evaluate the differential coefficients $f'(m), f''(m), \ldots$, it is natural to differentiate the series term by term and to add the results. This procedure would be adopted automatically for a series with a finite number of terms, and the assumption is that it also works for an infinite series. This assumption is not always justified, but it can be shown to hold in this comparatively simple case. Thus, successively,

$$f'(m) = a_1 + 2a_2 m + 3a_3 m^2 + \ldots + pa_p m^{p-1} + \ldots,$$
$$f''(m) = 2.1a_2 + 3.2a_3 m + \ldots + p(p-1)a_p m^{p-2} + \ldots,$$
$$f'''(m) = 3.2.1a_3 + \ldots + p(p-1)(p-2)a_p m^{p-3} + \ldots,$$

....

$$f^{(p)}(m) = p(p-1)\ldots 2.1a_p + \ldots,$$

and so on, so that, when m is zero,

$$f'(0) = a_1, \quad f''(0) = 2!a_2, \quad f'''(0) = 3!a_3, \quad \ldots, \quad f^{(p)}(0) = p!a_p, \quad \ldots,$$

or $$a_1 = k, \quad a_2 = \frac{k^2}{2!}, \quad a_3 = \frac{k^3}{3!}, \quad \ldots, \quad a_p = \frac{k^p}{p!}, \quad \ldots.$$

Hence, under these assumptions, the function $f(m)$ is obtained as an infinite series in the form

$$f(m) = 1+km+\frac{k^2m^2}{2!}+\frac{k^3m^3}{3!}+...+\frac{k^pm^p}{p!}+....$$

Now $f(m)$ was defined as an expression for a^m, so that a^m *has been obtained as an infinite series in the form*

$$a^m = 1+km+\frac{k^2m^2}{2!}+...+\frac{k^pm^p}{p!}+...,$$

where k, or $f'(0)$, is equal to the differential coefficient (with respect to m) of a^m, evaluated at $m = 0$.

The value of k depends on that of a. What we are looking for in this section is a value of a which will make the manipulations as simple as possible, and for that the series should be kept free of unnecessary complications. This suggests, if possible, a particular value of a for which $k = 1$. That particular value may be named by the next vowel, e, so that

$$e^m = 1+m+\frac{m^2}{2!}+\frac{m^3}{3!}+...+\frac{m^p}{p!}+....$$

The value of e itself is now found, in the form of an infinite series, by putting $m = 1$ (provided that the series converges). Thus

$$e = 1+1+\frac{1}{2!}+\frac{1}{3!}+...+\frac{1}{p!}+....$$

This series is known as the *exponential series*, and the theorem that, for this value of e,

$$e^m = 1+m+\frac{m^2}{2!}+...$$

is known as the *exponential theorem* governing the behaviour of powers of e with exponent m. Thus

$$\left(1+1+\frac{1}{2!}+\frac{1}{3!}+\frac{1}{4!}+...\right)^m = 1+m+\frac{m^2}{2!}+\frac{m^3}{3!}+\frac{m^4}{4!}+....$$

We must, for convergence, establish that, if S_n is the sum

$$S_n \equiv 1+1+\frac{1}{2!}+\frac{1}{3!}+...+\frac{1}{n!},$$

then S_n tends to a limit as $n \to \infty$; this limit is the number which we have called e. The sequence $S_1, S_2, S_3, S_4, ...$ increases steadily, each term being obtained from its predecessor by the addition of a positive quantity, and it is reasonable to assume (what can be proved rigorously) that *an increasing sequence which remains less than some ascertainable number must tend to a limit.* Now

$$\frac{1}{3.2} < \frac{1}{2.2} \quad \frac{1}{4.3.2} < \frac{1}{2.2.2},$$

and, more generally, $$\frac{1}{n!} < \frac{1}{2^{n-1}}.$$

Thus $$S_n < 1+1+\frac{1}{2}+\frac{1}{2^2}+\ldots+\frac{1}{2^{n-1}},$$

so that S_n is less than the sum of the infinite series

$$1+1+\frac{1}{2}+\frac{1}{2^2}+\frac{1}{2^3}+\ldots,$$

or $$1+2.$$

Hence, whatever the value of n,

$$S_n < 3,$$

and so S_n tends to a limit e, whose value is less than 3.

By evaluation of terms,

$$\begin{aligned} e = {} & 1 \\ & +1 \\ & +0\cdot5 \\ & +0\cdot166666\ldots \\ & +0\cdot041666\ldots \\ & +0\cdot008333\ldots \\ & +0\cdot001388\ldots \\ & +0\cdot000199\ldots \\ & +0\cdot000024\ldots \\ & \ldots\ldots\ldots\ldots\ldots\ldots \\ \hline e = {} & 2\cdot718\,(276\ldots). \end{aligned}$$

To four places of decimals, $e = 2\cdot7183$.

*3. Convergence for the series for e^m.

We prove that, *if*

$$S_n \equiv 1+m+\frac{m^2}{2!}+\ldots+\frac{m^n}{n!},$$

then S_n tends to a limit for all values of m.

* To be omitted or skimmed lightly at first reading.

Suppose first that m is positive. Let N be the first integer greater than m, and consider values of n greater than N. Then

$$S_n \equiv 1+m+\ldots+\frac{m^N}{N!}+\frac{m^{N+1}}{(N+1)!}+\ldots+\frac{m^n}{n!},$$

so that $$S_n-S_N = \frac{m^{N+1}}{(N+1)!}+\ldots+\frac{m^n}{n!}$$

$$= \frac{m^{N-1}}{(N+1)!}\left\{1+\frac{m}{N+2}+\frac{m^2}{(N+2)(N+3)}+\ldots\right\},$$

where there are $n-N$ terms in the bracket. Thus

$$S_n-S_N < \frac{m^{N+1}}{(N+1)!}\left\{1+\frac{m}{N}+\frac{m^2}{N^2}+\ldots\right\},$$

where the inequality remains true if the number of terms in the bracket is increased indefinitely. Thus, summing the infinite geometric series, we have the inequality

$$S_n-S_N < \frac{m^{N+1}}{(N+1)!}\,\frac{1}{1-(m/N)} = \frac{Nm^{N+1}}{(N+1)!(N-m)}.$$

Hence $$S_n < S_N+\frac{Nm^{N+1}}{(N+1)!(N-m)}.$$

Now the right-hand side is a definite fixed number, independent of n (since m is given and N is merely the first integer greater than it). Hence the sequence $S_1, S_2, S_3, S_4, \ldots$ of steadily increasing terms is always less than an ascertainable number, and so it tends to a limit (cf. p. 225).

Suppose next that m is negative, and write $m = -q$, where q is positive. Then

$$S_n = 1-q+\frac{q^2}{2!}-\ldots+(-1)^n\frac{q^n}{n!}.$$

Let N be the first *even* integer greater than q, and consider values of n greater than N. Then

$$S_n = \left(1-q+\frac{q^2}{2!}-\ldots+\frac{q^N}{N!}\right)-\frac{q^{N+1}}{(N+1)!}+\ldots+(-1)^n\frac{q^n}{n!},$$

so that $$S_N-S_n = \frac{q^{N+1}}{(N+1)!}-\frac{q^{N+2}}{(N+2)!}+\ldots-(-1)^n\frac{q^n}{n!}$$

$$= \frac{q^{N+1}}{(N+1)!}\left(1-\frac{q}{N+2}\right)+\frac{q^{N+3}}{(N+3)!}\left(1-\frac{q}{N+4}\right)+\ldots,$$

the series on the right terminating

(i) *if n is even,* with the term

$$\frac{q^{n-1}}{(n-1)!}\left(1-\frac{q}{n}\right),$$

(ii) *if n is odd,* with the terms

$$\frac{q^{n-2}}{(n-2)!}\left(1-\frac{q}{n-1}\right)+\frac{q^n}{n!}.$$

In either case all the terms on the right-hand side, as bracketed, are positive since $q < N$. Hence

$$S_n < S_N.$$

Now, once n is greater than N (and perhaps sooner) it is certainly true that the series

$$(1-q)+\left(\frac{q^2}{2!}-\frac{q^3}{3!}\right)+\ldots+\left(\frac{q^N}{N!}-\frac{q^{N+1}}{(N+1)!}\right)+\left(\frac{q^{N+2}}{(N+2)!}-\frac{q^{N+3}}{(N+3)!}\right)+\ldots$$

is steadily increasing, but it remains less than the definite number S_N. Hence (cf. p. 225) it tends to a limit. Thus S_n *tends to a limit for a sequence of odd values of n.* Call this limit S. Moreover, if n is *even,* then

$$S_n = (1-q)+\ldots+\left(\frac{q^{n-2}}{(n-2)!}-\frac{q^{n-1}}{(n-1)!}\right)+\frac{q^n}{n!};$$

and we now prove that the sum of the bracketed terms also tends to the limit S. It was proved (p. 202) that, as far as the 'extra' term is concerned,

$$\frac{q^n}{n!}\to 0.$$

Assuming as sufficiently obvious that the limit of a sum is equal to the sum of the limits, it therefore follows that

$$S_n \to S+0 = S.$$

Hence, whether n is even or odd,

$$S_n \to S,$$

so that *the series* $$1+m+\frac{m^2}{2!}+\frac{m^3}{3!}+\ldots$$

converges for all values of m, positive or negative.

4. Statement* of the exponential series. We have proved that there exists a number e, defined by the infinite series

$$e = 1+1+\frac{1}{2!}+\frac{1}{3!}+\ldots+\frac{1}{p!}+\ldots,$$

with the remarkable property that the powers of e are given by the relation

$$e^m = 1+m+\frac{m^2}{2!}+\frac{m^3}{3!}+\ldots+\frac{m^p}{p!}+\ldots,$$

the infinite series converging for *all* values of m. Just as

$$e^m \times e^n = e^{m+n},$$

so, correspondingly,

$$\left(1+m+\frac{m^2}{2!}+\ldots\right)\times\left(1+n+\frac{n^2}{2!}+\ldots\right) = 1+(m+n)+\frac{(m+n)^2}{2!}+\ldots.$$

Note that any positive number a can be expressed as a power of e in the form

$$a = e^k,$$

since, by taking logarithms to the base e of both sides, the actual value for k is obtained in the form

$$k = \log_e a.$$

Thus
$$a^m = 1+(\log_e a)m+\frac{(\log_e a)^2 m^2}{2!}+\frac{(\log_e a)^3 m^3}{3!}+\ldots.$$

Replacing m in the series for e^m by the more apparently variable letter x, we have the exponential series

$$e^x = 1+x+\frac{x^2}{2!}+\frac{x^3}{3!}+\ldots,$$

convergent for all values of the variable x, and we observe the important property (p. 222) that (subject to proper examination of convergence)

$$\frac{d}{dx}(e^x) = e^x,$$

so that *the function e^x is its own differential coefficient.*

Examples 1

Write out the first four terms and the kth term for each of the following expansions:

1. e^{2x}.
2. e^{-2x}.
3. e^{5x}.
4. e^{-x^2}.
5. e^{3y}.
6. e^{-4a}.

* This brief summary may be sufficient for the beginner, who should, however, return to the preceding work as soon as he is able.

7. $(1+x)e^x$. 8. $(1-x)e^{-x}$. 9. e^x+e^{2x}.

10. Prove that

(i) $e^x+e^{-x} = 2\left(1+\frac{x^2}{2!}+\frac{x^4}{4!}+\frac{x^6}{6!}+\ldots\right)$,

(ii) $e^x-e^{-x} = 2\left(x+\frac{x^3}{3!}+\frac{x^5}{5!}+\frac{x^7}{7!}+\ldots\right)$.

11. Find in terms of e the sums of the following series:

(i) $1+\frac{2}{2!}+\frac{3}{3!}+\frac{4}{4!}+\frac{5}{5!}+\ldots$,

(ii) $2+\frac{3}{2!}+\frac{4}{3!}+\frac{5}{4!}+\frac{6}{5!}+\ldots$,

(iii) $1+\frac{3}{2!}+\frac{5}{3!}+\frac{7}{4!}+\frac{9}{5!}+\ldots$.

12. Find the first four terms in each of the expansions

(i) e^{x+x^2}; (ii) e^{-x+x^2}; (iii) $e^{x+x^2+x^3}$.

Revision Examples XIII

1. If the product of e^x and the polynomial $1-x+\frac{1}{2}x^2-\frac{1}{6}x^3$ is written as a series in ascending powers of x, show that it contains no terms in x, x^2, x^3, and find (and simplify) the coefficient of x^n in this series when $n \geqslant 4$.

Deduce that, for all real values of x,

$$e^{-x} > 1-x+\frac{x^2}{2!}-\frac{x^3}{3!}.$$

2. Show that numbers a, b, c can be found such that the difference between

$$e^x(1-ax+bx^2-cx^3) \quad \text{and} \quad 1+ax+bx^2+cx^3$$

contains no power of x below the seventh; find a, b, c.

3. Evaluate as accurately as your tables allow the greatest term of the exponential series for $e^{8\cdot 1}$.

By comparison with an infinite geometrical progression, prove that the sum of all the terms of the series which follow the greatest term is less than

$$\frac{(8\cdot 1)^9\,100}{9!\,19}.$$

4. The expansion of $(a+bx+cx^2)e^{-x}$ begins with the terms

$$4-9x+8x^2.$$

Find a, b, c and show that the coefficient of x^n is $(-1)^n(n+2)^2/n!$

5. In the series
$$\frac{x}{1!}+\frac{x^2}{2!}+\frac{x^3}{3!}+\ldots$$
in which x is positive, u_n denotes the nth term. If this is the greatest term, prove that $n < x < n+1$.

If, in addition, $u_{n+1} < u_{n-1}$, find the values between which x must lie.

Prove that
$$x+\frac{x(x+3)}{1!}+\frac{x^2(x+6)}{2!}+\frac{x^3(x+9)}{3!}+\ldots = 4xe^x.$$

6. Find the coefficient of x^n in the expansion of the function

$$(1+px+qx^2)e^{-2x}.$$

7. Find the expansion of
$$e^{x-\frac{1}{2}x^2+\frac{1}{3}x^3-\frac{1}{4}x^4}$$
correct to x^5.

8. Find the coefficient of x^r in the expansion of

$$(1+3x)e^{-3x}$$

as a series of ascending powers of x.

9. Prove that

$$\frac{2^3}{2!}+\frac{3^3}{3!}+\frac{4^3}{4!}+\ldots \text{ to infinity } = 5e-1,$$

and find
$$\sum_{n=1}^{\infty}\frac{n^4}{n!}.$$

10. Find
$$\sum_{n=1}^{\infty}\frac{n-1}{(n+2)n!}x^n.$$

19

THE LOGARITHMIC SERIES

[As in the previous chapter, the preliminary discussion may be found hard. It should, however, be read lightly at least, so that the underlying structure may be seen.]

1. The logarithmic series. Before attempting to express the logarithm of a number in the form of an infinite series, observe that the function

$$\log x$$

has no meaning for $x = 0$, so that a series of the type

$$\log x = a_0 + a_1 x + a_2 x^2 + \ldots$$

would lead straight away to trouble for the value $x = 0$. On the other hand, $\log 1 = 0$ for all bases of logarithms, so it is worth while to try instead to form an expansion

$$\log_e(1+x) = a_0 + a_1 x + a_2 x^2 + a_3 x^3 + \ldots,$$

taking the number e as the 'natural' base at which we arrived in the preceding chapter.

To obtain some idea of what to expect, write

$$u = \log_e(1+x),$$

so that

$$e^u = 1+x.$$

Differentiate* with respect to x. Then (p. 228)

$$e^u \frac{du}{dx} = 1,$$

or

$$\frac{du}{dx} = \frac{1}{e^u}$$

$$= \frac{1}{1+x}.$$

* This may involve 'argument in a circle', according to the way in which the whole treatment has proceeded in the calculus course.

We may therefore expect the solution (and shall shortly justify that expectation)

$$\text{‘}\frac{du}{dx} = 1-x+x^2-x^3+\ldots,$$

so that
$$u = A+x-\tfrac{1}{2}x^2+\tfrac{1}{3}x^3-\tfrac{1}{4}x^4+\ldots.$$

But $u = 0$ when $x = 0$; thus $A = 0$, and so

$$\log_e(1+x) = x-\tfrac{1}{2}x^2+\tfrac{1}{3}x^3-\tfrac{1}{4}x^4+\ldots.\text{’}$$

The argument may be justified as follows:

Write y for the difference between $\log_e(1+x)$ and the first $2p+1$ terms of the proposed series:

$$y \equiv \log_e(1+x)-\left\{x-\tfrac{1}{2}x^2+\tfrac{1}{3}x^3-\ldots+\frac{1}{2p+1}x^{2p+1}\right\}.$$

Then $y = 0$ when $x = 0$.

Also, by direct differentiation,

$$\frac{dy}{dx} = \frac{1}{1+x}-(1-x+x^2-\ldots+x^{2p})$$

$$\equiv \frac{-x^{2p+1}}{1+x}.$$

The function y has therefore a stationary value when $x = 0$. Moreover, dy/dx passes from $(+)$ through 0 to $(-)$ as x passes from $(-)$ through 0 to $(+)$, and so the turning value is a maximum. Since $y = 0$ at this, the only, maximum, it follows that y can never be positive. Hence

$$y \leqslant 0,$$

so that
$$\log_e(1+x) \leqslant x-\tfrac{1}{2}x^2+\tfrac{1}{3}x^3-\ldots+\frac{1}{2p+1}x^{2p+1}.$$

Next, write, similarly,

$$z = \log_e(1+x)-\left\{x-\tfrac{1}{2}x^2+\tfrac{1}{3}x^3-\ldots-\frac{1}{2p}x^{2p}\right\}.$$

Then $z = 0$ when $x = 0$.

Also
$$\frac{dz}{dx} = \frac{1}{1+x}-(1-x+x^2-\ldots-x^{2p-1})$$

$$= \frac{x^{2p}}{1+x}.$$

We confine our attention to values of x such that

$$x > -1,$$

thereby ensuring that the denominator $1+x$ is positive. Then dz/dx is always positive, so that the function z increases steadily with x. Since $z = 0$ when $x = 0$, it follows that

$$z \geqslant 0 \quad \text{when} \quad x \geqslant 0,$$

$$z \leqslant 0 \quad \text{when} \quad x \leqslant 0,$$

with $z = 0$ only at $x = 0$.

Thus $$\log_e(1+x) \geqslant x - \tfrac{1}{2}x^2 + \tfrac{1}{3}x^3 - \ldots - \frac{1}{2p}x^{2p} \quad (x \geqslant 0),$$

$$\log_e(1+x) \leqslant x - \tfrac{1}{2}x^2 + \tfrac{1}{3}x^3 - \ldots - \frac{1}{2p}x^{2p} \quad (x \leqslant 0).$$

(i) *Positive values of* x. We have proved that

$$x - \tfrac{1}{2}x^2 + \ldots - \frac{1}{2p}x^{2p} \leqslant \log_e(1+x) \leqslant x - \tfrac{1}{2}x^2 + \ldots + \frac{1}{2p+1}x^{2p+1}.$$

The difference between the two series is, numerically,

$$\frac{x^{2p+1}}{2p+1},$$

and this tends to 0 as $p \to \infty$ for $x \leqslant 1$.

[On the other hand, if $x > 1$, say $x = 1+k$, with $k > 0$, then

$$\frac{(1+k)^{2p+1}}{2p+1} = \frac{1+(2p+1)k+\tfrac{1}{2}[(2p+1)2p]k^2+\ldots}{2p+1}$$

$$> \frac{1}{2p+1} + k + \ldots$$

$$> k,$$

so that $x^{2p+1}/(2p+1)$ cannot tend to zero.]

Hence, if $0 \leqslant x \leqslant 1$, the two series tend to equality as $p \to \infty$, and the value to which they tend must be $\log_e(1+x)$ lying between them. Thus we have established the relation

$$\log_e(1+x) = x - \tfrac{1}{2}x^2 + \tfrac{1}{3}x^3 - \tfrac{1}{4}x^4 + \ldots$$

for $$0 \leqslant x \leqslant 1.$$

(ii) *Negative* values of* x. We have proved that

$$\log_e(1+x) \leqslant x - \tfrac{1}{2}x^2 + \ldots + \frac{1}{2p+1}x^{2p+1}$$

* This is hard, and should certainly be omitted at first.

and that

$$\log_e(1+x) \leqslant x - \tfrac{1}{2}x^2 + \ldots - \frac{1}{2p}x^{2p},$$

the latter inequality under the condition $x > -1$. Unfortunately both of these inequalities are in the same sense, so that we cannot use the method exactly as before.

To avoid constant use of negative numbers, write $x = -u$, where $0 \leqslant u < 1$. The two inequalities give the relation (for a number of terms which is either odd or even)

$$\log_e(1-u) \leqslant -u - \tfrac{1}{2}u^2 - \tfrac{1}{3}u^3 - \ldots - \frac{1}{n}u^n,$$

or

$$u + \tfrac{1}{2}u^2 + \tfrac{1}{3}u^3 + \ldots + \frac{1}{n}u^n \leqslant \log \frac{1}{1-u},$$

whether n is odd or even.

Consider now the function

$$w \equiv u + \tfrac{1}{2}u^2 + \ldots + \frac{1}{n-1}u^{n-1} + \frac{1}{n}u^n + u^n - \log \frac{1}{1-u}.$$

Then $w = 0$ when $u = 0$. Also

$$\frac{dw}{du} = 1 + u + \ldots + u^{n-2} + u^{n-1} + nu^{n-1} - \frac{1}{1-u}$$

$$= \frac{1-u^n}{1-u} + nu^{n-1} - \frac{1}{1-u} = nu^{n-1} - \frac{u^n}{1-u}.$$

Now this differential coefficient is positive provided that $n > u/(1-u)$, and we choose n sufficiently large for this inequality to hold.* Then dw/du is positive for $0 \leqslant u < 1$. Hence w is an increasing function of u in that interval, and, since $w = 0$ when $u = 0$, it follows that

$$w \geqslant 0 \quad (0 \leqslant u < 1).$$

Hence (replacing u by $-x$ in the expression for w)

$$-x + \tfrac{1}{2}x^2 - \ldots + (-1)^n \frac{1}{n}x^n + (-1)^n x^n + \log_e(1+x) \geqslant 0,$$

so that

$$\log_e(1+x) \geqslant x - \tfrac{1}{2}x^2 + \ldots + (-1)^{n-1}\frac{1}{n}x^n + (-1)^{n-1}x^n.$$

* More correctly, we ought to have *fixed* a number δ such that $0 < \delta < 1$ and then taken $0 \leqslant u \leqslant \delta$. The number n can then be chosen to exceed the fixed number $\delta/(1-\delta)$, and so, necessarily, to exceed $u/(1-u)$. This ensures that n does not become unbounded in an approach of u to 1.

Combination of this with the earlier relation gives the inequalities

$$x-\tfrac{1}{2}x^2+\ldots+(-1)^{n-1}\frac{1}{n}x^n+(-1)^{n-1}x^n$$
$$\leqslant \log_e(1+x) \leqslant x-\tfrac{1}{2}x^2+\ldots+(-1)^n\frac{1}{n}x^n.$$

Now the right-hand series exceeds the left-hand by

$$-(-1)^{n-1}x^n,$$

or

$$(-x)^n,$$

a positive quantity. Also, since $|x| < 1$,

$$(-x)^n \to 0.$$

Hence the two series tend to equality, and $\log_e(1+x)$, lying between them, is the limit of either. Hence

$$\log_e(1+x) = x-\tfrac{1}{2}x^2+\tfrac{1}{3}x^3-\tfrac{1}{4}x^4+\ldots,$$

for

$$-1 < x \leqslant 0.$$

SUMMARY. *The function* $\log_e(1+x)$ *may be expressed as an infinite series in the form*

$$\log_e(1+x) = x-\tfrac{1}{2}x^2+\tfrac{1}{3}x^3-\tfrac{1}{4}x^4+\ldots$$

for values of x *in the interval* $-1 < x \leqslant 1$.

***Illustration* 1.** *To prove that, if* $n > 1$, *then*

$$2\log_e n-\log_e(n+1)-\log_e(n-1) = \frac{1}{n^2}+\frac{1}{2n^4}+\frac{1}{3n^6}+\ldots,$$

and hence, given that

$$\log_e 10 = 2{\cdot}30259,\quad \log_e 3 = 1{\cdot}09861,$$

to calculate $\log_e 11$ *to four places of decimals.*

Expressed as a single logarithm,

$$2\log_e n-\log_e(n+1)-\log_e(n-1)$$
$$= \log_e\frac{n^2}{(n+1)(n-1)} = \log_e\frac{n^2}{n^2-1}$$
$$= \log_e\frac{1}{1-(1/n^2)} = -\log_e(1-n^{-2}).$$

Hence, using the logarithmic series, we have

$$-\left\{\frac{-1}{n^2}-\frac{1}{2}\left(\frac{-1}{n^2}\right)^2+\frac{1}{3}\left(\frac{-1}{n^2}\right)^3-\ldots\right\} = \frac{1}{n^2}+\frac{1}{2n^4}+\frac{1}{3n^6}+\ldots.$$

This expression is valid so long as

$$-1 < \frac{-1}{n^2} \leqslant 1,$$

so that $$n^2 > 1,$$

or, assuming n to be positive, $$n > 1.$$

Now put $n = 10$. Then

$$2\log_e 10 - \log_e 11 - \log_e 9 = \tfrac{1}{100} + \tfrac{1}{20\,000} + \tfrac{1}{3\,000\,000} + \ldots.$$

Hence

$$\begin{aligned}\log_e 11 &= 2(\log_e 10 - \log_e 3) - 0{\cdot}01 - \tfrac{1}{2}(0{\cdot}0001) - \tfrac{1}{3}(0{\cdot}000001) - \ldots\\ &= 2{\cdot}40796 - 0{\cdot}01 - 0{\cdot}00005 - 0{\cdot}0000003\ldots - \ldots = 2{\cdot}3979(1),\end{aligned}$$

so that to four places of decimals,

$$\log_e 11 = 2{\cdot}3979.$$

Examples 1

1. Expand in ascending powers of x and state the ranges of values of x for which the expansions are valid:

(i) $\log_e(1+2x)$. (ii) $\log_e(1-3x)$.

(iii) $\log_e(1+x^2)$. (iv) $\log_e(1+3x+2x^2)$.

(v) $\log_e \dfrac{1+x}{1-x}$. (vi) $\log_e \dfrac{1+3x}{1+2x}$.

2. Prove that

$$\log_e(2+x) = \log_e 2 + \tfrac{1}{2}x - \tfrac{1}{2}(\tfrac{1}{2}x)^2 + \tfrac{1}{3}(\tfrac{1}{2}x)^3 - \tfrac{1}{4}(\tfrac{1}{2}x)^4 + \ldots,$$

and expand similarly the following expressions

(i) $\log_e(2+3x)$. (ii) $\log_e(2-3x)$.

(iii) $\log_e(9-4x^2)$. (iv) $\log_e \dfrac{2+x}{2-x}$.

(v) $\log_e(4x^2-5x+1)$. (vi) $\log_e(6-5x-6x^2)$.

Revision Examples XIV

1. Expand $$\log_e \frac{2+x}{1-x}$$

in ascending powers of x, giving the coefficient of x^r and the numerical values of the first three terms.

2. Expand $$\log(6-5x)-\log 6$$
in a series of ascending powers of x.

Also expand $$\log\frac{1}{6-5x+x^2}$$
in a similar series, giving the coefficient of x^r.

State the values of x for which (i) the first series, (ii) the second series, is convergent.

3. Show that, if x is small, an approximate value of $\{\log_e(1+x)\}^n$ is
$$x^n-\tfrac{1}{2}nx^{n+1}+\tfrac{1}{24}n(3n+5)x^{n+2}.$$

4. Find the first three terms and the coefficient of x^r in the expansion of
$$\log_e(1-x-2x^2)$$
in ascending powers of x, and state the range of values of x for which the expansion is valid.

5. Expand $$\log_e\frac{(1-3x)^2}{(1-2x)^3}$$
as far as the term in x^4, and find the coefficient of x^r.

State the range of values of x for which the expansion is valid.

6. Find the coefficient of x^r in the expansion of
$$(1+3x)^2e^{-6x},$$
and prove that the logarithm, to the base e, of this expression is
$$-9x^2+18x^3$$
as far as the term in x^3.

7. Show that, if $n>1$,
$$\log_e\frac{n+1}{n-1}=2\left(\frac{1}{n}+\frac{1}{3n^3}+\frac{1}{5n^5}+\ldots\right),$$
and calculate $\log_e 9$ (as $2\log_e 3$) to 4 places of decimals.

8. Expand $\log_e\left(1+\frac{1}{n}\right)$ in ascending powers of $\frac{1}{2n+1}$, stating the necessary restrictions on the value of n.

Prove that, if n has any positive value, $\log_e\left(1+\frac{1}{n}\right)$ is greater than $\frac{2}{2n+1}$ and less than $\frac{2n+1}{2n(n+1)}$.

9. From the formula

$$\log_e\left(\frac{1+x}{1-x}\right) = 2(x+\tfrac{1}{3}x^3+\tfrac{1}{5}x^5+\ldots)$$

compute the value of $\log_e(1\cdot25)$ to 6 places of decimals.

10. Assuming that $n > 1$, prove that

$$\log_e\left(\frac{n}{n-1}\right) = \frac{1}{n}+\frac{1}{2n^2}+\frac{1}{3n^3}+\ldots,$$

and, by putting $n = 10$, find the value of $\log_{10}3$ to 4 places of decimals, given that $\log_{10}e = 0\cdot4343$.

11. Prove that, if $n > 2$,

$$\tfrac{1}{2}\log_{10}\left(\frac{n+2}{n-2}\right)-\log_{10}\left(\frac{n+1}{n-1}\right) = \mu\left(\frac{1}{z}+\frac{1}{3z^3}+\frac{1}{5z^5}+\ldots\right),$$

where $z = \tfrac{1}{2}(n^3-3n)$ and $\mu = \log_{10}e$.

Verify this result from the tables when $n = 6$, taking μ to be $0\cdot4343$.

12. If

$$y = \frac{x}{1+x},$$

where x is any positive number, show that $\log_e(1+x)$ may be expanded in ascending powers of y, obtaining the expansion.

Deduce that, for all positive values of x,

$$\log_e(1+x) > \frac{x}{1+x}.$$

13. Prove that

$$\log_e x = 2\left\{\frac{x-1}{x+1}+\frac{1}{3}\left(\frac{x-1}{x+1}\right)^3+\frac{1}{5}\left(\frac{x-1}{x+1}\right)^5+\ldots\right\}.$$

Find the value of $\log_e 2$ correct to 4 places of decimals.

Prove also that, approximately,

$$\log_e(\tfrac{1}{3}\sqrt{11}) = 0\cdot100335.$$

14. Define $\log_b N$, and deduce from the definition that

$$\log_e\frac{ay}{z^2} = \log_e a+\log_e y-2\log_e z.$$

Find the first four terms of the expansion of

$$\log_e\frac{2+x}{(1-x)^2}$$

in a series of ascending powers of x, and state the range of values of x for which the expansion is valid.

15. If $a = \log_e \frac{14}{15}$, $b = \log_e \frac{21}{20}$, $c = \log_e \frac{49}{50}$,

prove that
$$a+b-c = 0,$$
$$a+3b-2c = \log_e \tfrac{9}{8}.$$

Assuming that $\log_e(1+x) = x-\frac{1}{2}x^2+\frac{1}{3}x^3-\ldots,$

calculate the values of b, c to five places of decimals, and deduce the value of $\log_e \frac{15}{14}$ and $\log_e \frac{9}{8}$ to four places of decimals.

16. If
$$a = -\log(1-\tfrac{1}{10}),\quad b = -\log(1-\tfrac{4}{100}),\quad c = \log(1+\tfrac{1}{80}),$$
show that
$$\log 2 = 7a-2b+3c,$$
$$\log 5 = 16a-4b+7c.$$

Hence, by using the logarithmic series, calculate $\log_e 10$ to four significant figures.

17. Prove that, when x is sufficiently small,
$$\frac{x}{\log(1+x)} = 1+\tfrac{1}{2}x-\tfrac{1}{12}x^2+\tfrac{1}{24}x^3+\ldots.$$

18. Express
$$\frac{(4-x)^2}{4(2+x)^2(1-x)}$$
in partial fractions, and show, with the help of the logarithmic series, that, for $0 < x < 1$,
$$\frac{x(4-x)^2}{4(2+x)^2(1-x)} > \log_e(1+x).$$

19. Show that, if
$$\log_e y = 1+\tfrac{1}{2}x-\tfrac{1}{6}x^2+12x^3,$$
then, as far as the third power in x,
$$y = e(1+\tfrac{1}{2}x-\tfrac{1}{24}x^2+\tfrac{1}{48}x^3).$$

20. By taking logarithms, or otherwise, verify that, when n is large, an approximate value of $(1+(1/n))^n$ is
$$e\left(1-\frac{1}{2n}+\frac{11}{24n^2}-\frac{7}{16n^3}\right).$$

21. The expansion in ascending powers of x of the expression
$$e^{-5x}-ke^{x}+h(1+x)^h$$
is $a+bx+cx^2+\ldots$. Determine positive integers h, k so that a, b are both zero.

22. Find the value of the first term which does not vanish in the expansion of

$$2e^{3x}+2e^{x}-2e^{-x}+11(1+x)^3-22(1+x)^4+9(1+x)^5$$

in ascending powers of x.

23. Find the first non-vanishing term in the expansion of

$$(1-e^x)(1+\tfrac{1}{3}x)^{-3}+\log_e(1+x).$$

24. It is given that y is the positive value of

$$(1+x+x^2)^{1/x^2}.$$

By means of the expansions of $\log_e(1+x)$ and e^x, prove that, when x is small,

$$y = e^{(x+2)/2x}(1-\tfrac{2}{3}x+\ldots).$$

25. Prove that

$$\frac{e-1}{e+1}+\frac{1}{3}\left(\frac{e-1}{e+1}\right)^3+\frac{1}{5}\left(\frac{e-1}{e+1}\right)^5+\ldots = \frac{1}{2}.$$

26. Prove that, when x is sufficiently small, numerically,

$$(1+2x)^{-\frac{1}{2}}e^x = 1+x^2-\tfrac{4}{3}x^3+\ldots$$

and find the coefficient of x^4.

For what values of x is the expansion valid?

27. Find the values of a, b when the expansion of e^x in powers of x is identical with the expansion of

$$\frac{1+ax}{1-bx}$$

as far as the term in x^2.

Deduce that with these values of a, b

$$\frac{1+ax}{1-bx}-e^x = \tfrac{1}{12}x^3$$

when x is so small that x^4 can be neglected.

28. Determine a, b so that the expansion of

$$\frac{1+ax}{1+bx}\log_e(1+x)$$

may contain no terms in x^2 or x^3, and show that with these values

$$\frac{1+bx}{1+ax} = 1-\tfrac{1}{2}x+\tfrac{1}{3}x^2-\tfrac{2}{9}x^3,$$

neglecting powers of x above the third.

29. Prove that

$$\frac{e^{x}+e^{-x}}{e^{3x}} = 2-6x+20\frac{x^2}{2!}-\ldots+(-1)^n(4^n+2^n)\frac{x^n}{n!}+\ldots.$$

30. The function $$e^{-x}-\frac{1-x}{(1-x^2)^{\frac{1}{2}}(1-x^3)^{\frac{1}{3}}}$$

is expanded in ascending powers of x in the form

$$ax+bx^2+cx^3+dx^4+\ldots.$$

Prove that a, b, c, d are all zero.

31. Find the expansion in powers of x of $e^{x}-e^{-x}$, and, if $x = 1+kt$, where t is small, show that, correct to the square of t,

$$e^{x}-e^{-x} = \left(e-\frac{1}{e}\right)+kt\left(e+\frac{1}{e}\right)+\tfrac{1}{2}k^2t^2\left(e-\frac{1}{e}\right).$$

20

ELEMENTARY PROPERTIES OF DETERMINANTS

1. The solution of two linear equations. Let

$$a_1x+b_1y+c_1 = 0,$$
$$a_2x+b_2y+c_2 = 0$$

be two simultaneous equations in x, y. In accordance with the normal procedure for solution, multiply by b_2, b_1 and subtract:

$$(a_1b_2-a_2b_1)x+(c_1b_2-c_2b_1) = 0;$$

multiply also by a_2, a_1 and subtract:

$$(b_1a_2-b_2a_1)y+(c_1a_2-c_2a_1) = 0.$$

If the coefficients are perfectly general, so that $a_1b_2-a_2b_1 \neq 0$, the solution is

$$x = \frac{b_1c_2-b_2c_1}{a_1b_2-a_2b_1}, \quad y = -\frac{a_1c_2-a_2c_1}{a_1b_2-a_2b_1}.$$

It is sometimes found convenient to exhibit this solution in ratio form:

$$\frac{x}{b_1c_2-b_2c_1} = \frac{-y}{a_1c_2-a_2c_1} = \frac{1}{a_1b_2-a_2b_1}.$$

2. The determinantal structure. The three expressions just obtained in the denominators can be constructed by a simple rule. The coefficients in the two given equations are written in 'block' form:

$$\begin{matrix} a_1 & b_1 & c_1 \\ a_2 & b_2 & c_2. \end{matrix}$$

The elements used for the first term $b_1c_2-b_2c_1$ of the solution are found in the square formed by omitting the first column of the 'block'; those for the second term $a_1c_2-a_2c_1$ are in the square formed by omitting the second column; and those for the third term $a_1b_2-a_2b_1$ are in the square formed by omitting the third column. This observation supplies the basis for a notation and a rule of calculation.

A notation for the three expressions may be obtained by omitting the relevant column from the 'block' and then enclosing the resulting square between vertical lines. Thus we write

$$\begin{vmatrix} b_1 & c_1 \\ b_2 & c_2 \end{vmatrix} \equiv b_1 c_2 - b_2 c_1,$$

$$\begin{vmatrix} a_1 & c_1 \\ a_2 & c_2 \end{vmatrix} \equiv a_1 c_2 - a_2 c_1,$$

$$\begin{vmatrix} a_1 & b_1 \\ a_2 & b_2 \end{vmatrix} \equiv a_1 b_2 - a_2 b_1.$$

The three expressions on the left are called *determinants*.

The values of these determinants may be calculated by the rule, illustrated for the first of them:

In the determinant

$$\begin{vmatrix} b_1 & c_1 \\ b_2 & c_2 \end{vmatrix},$$

(i) multiply the two elements b_1, c_2 in the diagonal of sense $\diagdown$ and take the product with *positive* sign, giving $+b_1 c_2$;

(ii) multiply the two elements c_1, b_2 in the diagonal of sense $\diagup$ and take the product with *negative* sign, giving $-b_2 c_1$.

In the notation of determinants, the solution of the two given equations is

$$\frac{x}{\begin{vmatrix} b_1 & c_1 \\ b_2 & c_2 \end{vmatrix}} = \frac{-y}{\begin{vmatrix} a_1 & c_1 \\ a_2 & c_2 \end{vmatrix}} = \frac{1}{\begin{vmatrix} a_1 & b_1 \\ a_2 & b_2 \end{vmatrix}}.$$

The negative sign in front of y is very important.

One or two definitions may be given, in terms of the determinant

$$\begin{vmatrix} b_1 & c_1 \\ b_2 & c_2 \end{vmatrix}.$$

The numbers b_1, c_1, b_2, c_2 are called *elements*; the elements b_1, c_1 and b_2, c_2 are said to be in *rows* and the elements b_1, b_2 and c_1, c_2 in *columns*. The elements b_1, c_2 lie in the *leading diagonal*; their product appears in the expansion $b_1 c_2 - b_2 c_1$ with positive sign.

***Illustration* 1.** (i) *To prove that*

$$\begin{vmatrix} \cos\theta & \sin\theta \\ -\sin\theta & \cos\theta \end{vmatrix} = 1.$$

The left-hand side is

$$\cos\theta . \cos\theta - \sin\theta\,(-\sin\theta) = \cos^2\theta + \sin^2\theta = 1.$$

(ii) *To prove that*

$$\begin{vmatrix} 1 & i \\ -i & 1 \end{vmatrix} = 0.$$

where $i^2 = -1$.

The left-hand side is

$$\begin{aligned} 1-(-i)i &= 1+i^2 \\ &= 0. \end{aligned}$$

3. **Elimination.** The determinantal form can be reached from an alternative, though related, point of view. Consider the two equations

$$\begin{aligned} a_1x+b_1 &= 0, \\ a_2x+b_2 &= 0. \end{aligned}$$

If $a_1 \neq 0$, the first equation can be solved to give

$$x = -b_1/a_1,$$

but this solution does not generally satisfy the second equation. The condition for it to do so is

$$a_2(-b_1/a_1)+b_2 = 0,$$

or, multiplying by a_1 in order to express the left-hand side as a polynomial in the coefficients,

$$a_1b_2-a_2b_1 = 0.$$

We recognize the determinantal form

$$\begin{vmatrix} a_1 & b_1 \\ a_2 & b_2 \end{vmatrix} = 0.$$

This is a relation from which the variable x has been *eliminated*, and the left-hand side, the determinant of the coefficients, is called the *eliminant* of the two equations. The vanishing of the eliminant is the condition for the two equations in the one variable x to hold simultaneously.

Further progress is made by proceeding next to the three equations

$$\begin{aligned} a_1x+b_1y+c_1 &= 0, \\ a_2x+b_2y+c_2 &= 0, \\ a_3x+b_3y+c_3 &= 0. \end{aligned}$$

The first two of them can, in general, be solved, by §2, in the form

$$\frac{x}{b_1c_2-b_2c_1} = \frac{-y}{a_1c_2-a_2c_1} = \frac{1}{a_1b_2-a_2b_1},$$

but this solution does not necessarily satisfy the third equation. The condition for it to do so is

$$a_3(b_1c_2-b_2c_1)-b_3(a_1c_2-a_2c_1)+c_3(a_1b_2-a_2b_1) = 0.$$

The left-hand side is the *eliminant* of the three equations.

This remark leads to a generalization of the concept of a determinant, which is the object of the next paragraph.

4. The determinantal structure generalized. The nine coefficients of the three equations at the end of the preceeding paragraph are, in 'block' form,

$$\begin{matrix} a_1 & b_1 & c_1 \\ a_2 & b_2 & c_2 \\ a_3 & b_3 & c_3. \end{matrix}$$

The eliminant, re-written in the form

$$a_1(b_2c_3-b_3c_2)-b_1(a_2c_3-a_3c_2)+c_1(a_2b_3-a_3b_2),$$

can be derived in the following manner:

(i) Write down the coefficients in the first row:

$$a_1 \quad b_1 \quad c_1.$$

(ii) Write down the three determinants obtained from the remaining rows by omitting in turn the first, second and third columns, *and give them alternately the signs* $+, -, +$:

$$\begin{vmatrix} b_2 & c_2 \\ b_3 & c_3 \end{vmatrix}, \quad -\begin{vmatrix} a_2 & c_2 \\ a_3 & c_3 \end{vmatrix}, \quad \begin{vmatrix} a_2 & b_2 \\ a_3 & b_3 \end{vmatrix}.$$

(iii) Multiply corresponding elements from (i), (ii), and add:

$$a_1\begin{vmatrix} b_2 & c_2 \\ b_3 & c_3 \end{vmatrix} - b_1\begin{vmatrix} a_2 & c_2 \\ a_3 & c_3 \end{vmatrix} + c_1\begin{vmatrix} a_2 & b_2 \\ a_3 & b_3 \end{vmatrix}.$$

This, on expansion of the determinants, gives the eliminant.

The function

$$a_1(b_2c_3-b_3c_2)-b_1(a_2c_3-a_3c_2)+c_1(a_2b_3-a_3b_2)$$

is called a determinant *with three rows and columns*, and is written in the notation

$$\begin{vmatrix} a_1 & b_1 & c_1 \\ a_2 & b_2 & c_2 \\ a_3 & b_3 & c_3 \end{vmatrix}.$$

The function may also be written in two more symmetrical ways:

$$a_1(b_2c_3-b_3c_2)+b_1(c_2a_3-c_3a_2)+c_1(a_2b_3-a_3b_2),$$

$$a_1(b_2c_3-b_3c_2)+a_2(b_3c_1-b_1c_3)+a_3(b_1c_2-b_2c_1).$$

The expansion without brackets is

$$a_1b_2c_3-a_1b_3c_2+a_2b_3c_1-a_2b_1c_3+a_3b_1c_2-a_3b_2c_1.$$

Illustration 2. *To evaluate the determinant*

$$\begin{vmatrix} 1 & 3 & -5 \\ 2 & -4 & 6 \\ 5 & 7 & 1 \end{vmatrix}.$$

The value is

$$1\begin{vmatrix} -4 & 6 \\ 7 & 1 \end{vmatrix} - 3\begin{vmatrix} 2 & 6 \\ 5 & 1 \end{vmatrix} + (-5)\begin{vmatrix} 2 & -4 \\ 5 & 7 \end{vmatrix}$$

$$= (-4-42)-3(2-30)-5(14+20)$$

$$= -46+84-170$$

$$= -132.$$

Illustration 3. *To evaluate the determinant*

$$\begin{vmatrix} 3 & -2 & 4 \\ -4 & 1 & -7 \\ 5 & 3 & 13 \end{vmatrix}.$$

The value is

$$3\begin{vmatrix} 1 & -7 \\ 3 & 13 \end{vmatrix} - (-2)\begin{vmatrix} -4 & -7 \\ 5 & 13 \end{vmatrix} + 4\begin{vmatrix} -4 & 1 \\ 5 & 3 \end{vmatrix}$$

$$= 3(13+21)+2(-52+35)+4(-12-5)$$

$$= 102-34-68$$

$$= 0.$$

Since the value of this determinant is zero, we are led (cf. § 3) to consider the three equations

$$3x-2y+4 = 0,$$

$$-4x+y-7 = 0,$$

$$5x+3y+13 = 0.$$

The solution of the first two is

$$x = -2, \quad y = -1,$$

and these values also satisfy the third equation, as might have been expected. (But we are far from any real *proof* that this need happen in general.)

Illustration 4. To evaluate the determinant

$$\begin{vmatrix} 1 & 1 & 1 \\ a & b & c \\ a^2 & b^2 & c^2 \end{vmatrix}.$$

The value is $1\begin{vmatrix} b & c \\ b^2 & c^2 \end{vmatrix} - 1\begin{vmatrix} a & c \\ a^2 & c^2 \end{vmatrix} + 1\begin{vmatrix} a & b \\ a^2 & b^2 \end{vmatrix}$

$$= (bc^2-b^2c)-(ac^2-a^2c)+(ab^2-a^2b),$$

or, more symmetrically,

$$-\{bc(b-c)+ca(c-a)+ab(a-b)\}.$$

Examples 1

Evaluate the determinants:

1. $\begin{vmatrix} 2 & 5 \\ 1 & 3 \end{vmatrix}$.

2. $\begin{vmatrix} 4 & -2 \\ 6 & 8 \end{vmatrix}$.

3. $\begin{vmatrix} -2 & 7 \\ -1 & 3 \end{vmatrix}$.

4. $\begin{vmatrix} a & 1 \\ b & 1 \end{vmatrix}$.

5. $\begin{vmatrix} a & b \\ b & a \end{vmatrix}$.

6. $\begin{vmatrix} a & b \\ a & b \end{vmatrix}$.

7. $\begin{vmatrix} 3 & 2 & 4 \\ 1 & 5 & 6 \\ 2 & 1 & 3 \end{vmatrix}$.

8. $\begin{vmatrix} 5 & 4 & 1 \\ 2 & 3 & 4 \\ 6 & 2 & 3 \end{vmatrix}$.

9. $\begin{vmatrix} 3 & -2 & 4 \\ -5 & 1 & 6 \\ 7 & 8 & -1 \end{vmatrix}$.

10. $\begin{vmatrix} -4 & -2 & 6 \\ 8 & 3 & -1 \\ -9 & -7 & 5 \end{vmatrix}$.

11. $\begin{vmatrix} 1 & 2 & 3 \\ 2 & 3 & 4 \\ 3 & 4 & 5 \end{vmatrix}$.

12. $\begin{vmatrix} 5 & -3 & 1 \\ 2 & -4 & 6 \\ 9 & -11 & 13 \end{vmatrix}$.

13. $\begin{vmatrix} a & b & c \\ c & a & b \\ b & c & a \end{vmatrix}$.

14. $\begin{vmatrix} a & h & g \\ h & b & f \\ g & f & c \end{vmatrix}$.

15. $\begin{vmatrix} 1 & 0 & 0 \\ 3 & 5 & -6 \\ 2 & 4 & 3 \end{vmatrix}$.

16. $\begin{vmatrix} 0 & -2 & 3 \\ 5 & 7 & 6 \\ 1 & 0 & 9 \end{vmatrix}$.

17. $\begin{vmatrix} 0 & -h & g \\ h & 0 & -f \\ -g & f & 0 \end{vmatrix}$.

18. $\begin{vmatrix} a_1 & a_2 & a_3 \\ b_1 & b_2 & b_3 \\ c_1 & c_2 & c_3 \end{vmatrix}$.

5. Properties of determinants. These properties, though not the proofs, hold for determinants of any number of rows and columns.

(i) *The determinant is unaltered in value if rows and columns are interchanged.* That is,

$$\begin{vmatrix} a_1 & a_2 & a_3 \\ b_1 & b_2 & b_3 \\ c_1 & c_2 & c_3 \end{vmatrix} = \begin{vmatrix} a_1 & b_1 & c_1 \\ a_2 & b_2 & c_2 \\ a_3 & b_3 & c_3 \end{vmatrix}.$$

The left-hand side, by definition, is

$$a_1\begin{vmatrix} b_2 & b_3 \\ c_2 & c_3 \end{vmatrix} - a_2\begin{vmatrix} b_1 & b_3 \\ c_1 & c_3 \end{vmatrix} + a_3\begin{vmatrix} b_1 & b_2 \\ c_1 & c_2 \end{vmatrix}$$

$$= a_1b_2c_3 - a_1b_3c_2 - a_2b_1c_3 + a_2b_3c_1 + a_3b_1c_2 - a_3b_2c_1$$

= the right-hand side (p. 246).

(ii) *The determinant changes sign, keeping its numerical value, if two rows (columns) are interchanged.* That is, for example,

$$\begin{vmatrix} a_2 & b_2 & c_2 \\ a_1 & b_1 & c_1 \\ a_3 & b_3 & c_3 \end{vmatrix} = -\begin{vmatrix} a_1 & b_1 & c_1 \\ a_2 & b_2 & c_2 \\ a_3 & b_3 & c_3 \end{vmatrix}.$$

The left-hand side, by definition, is

$$a_2\begin{vmatrix} b_1 & c_1 \\ b_3 & c_3 \end{vmatrix} - b_2\begin{vmatrix} a_1 & c_1 \\ a_3 & c_3 \end{vmatrix} + c_2\begin{vmatrix} a_1 & b_1 \\ a_3 & b_3 \end{vmatrix}$$

$$= a_2b_1c_3 - a_2c_1b_3 - b_2a_1c_3 + b_2c_1a_3 + c_2a_1b_3 - c_2a_3b_1$$

= the right-hand side.

(iii) *The value of a determinant is zero if two rows (columns) are identical.* If the value is Δ, the interchange of the two identical rows (columns) gives, by (ii), a determinant of value $-\Delta$. But the interchange of what are identical does not affect the value. Hence

$$\Delta = -\Delta,$$

so that $$\Delta = 0.$$

(iv) *If all the elements of one row (column) are multiplied by k, the value of the determinant is multiplied by k.* That is, for example,

$$\begin{vmatrix} a_1 & b_1 & c_1 \\ ka_2 & kb_2 & kc_2 \\ a_3 & b_3 & c_3 \end{vmatrix} = k\begin{vmatrix} a_1 & b_1 & c_1 \\ a_2 & b_2 & c_2 \\ a_3 & b_3 & c_3 \end{vmatrix}.$$

The left-hand side is (cf. p. 246)

$$a_1(kb_2)c_3 - a_1 b_3(kc_2) + (ka_2)b_3 c_1 - (ka_2)b_1 c_3 + a_3 b_1(kc_2) - a_3(kb_2)c_1$$
$$= \text{the right-hand side.}$$

COROLLARY. *If all the elements of a determinant (of three rows and columns) are multiplied by k, then the value of the determinant is multiplied by k^3.*

(v) *If all the elements of one row (column) are expressed in composite form as a sum of the same number of terms, then the determinant can be expressed correspondingly as a sum of determinants.* That is, for example,

$$\begin{vmatrix} u_1+v_1+w_1 & u_2+v_2+w_2 & u_3+v_3+w_3 \\ b_1 & b_2 & b_3 \\ c_1 & c_2 & c_3 \end{vmatrix}$$
$$= \begin{vmatrix} u_1 & u_2 & u_3 \\ b_1 & b_2 & b_3 \\ c_1 & c_2 & c_3 \end{vmatrix} + \begin{vmatrix} v_1 & v_2 & v_3 \\ b_1 & b_2 & b_3 \\ c_1 & c_2 & c_3 \end{vmatrix} + \begin{vmatrix} w_1 & w_2 & w_3 \\ b_1 & b_2 & b_3 \\ c_1 & c_2 & c_3 \end{vmatrix}.$$

The left-hand side, by the expansion without brackets given in full on p. 246, is

$$(u_1+v_1+w_1)b_2 c_3 - (u_1+v_1+w_1)b_3 c_2 + \ldots$$
$$= (u_1 b_2 c_3 - u_1 b_3 c_2 + \ldots) + (v_1 b_2 c_3 - v_1 b_3 c_2 + \ldots) + (w_1 b_2 c_3 - w_1 b_3 c_2 + \ldots)$$
$$= \text{the right-hand side.}$$

COROLLARY. This result can be generalized widely. For example, a determinant such as

$$\begin{vmatrix} u_1+v_1+w_1 & u_2+v_2+w_2 & u_3+v_3+w_3 \\ p_1+q_1 & p_2+q_2 & p_3+q_3 \\ c_1 & c_2 & c_3 \end{vmatrix}$$

can be expressed as a sum of *six* determinants like

$$\begin{vmatrix} u_1 & u_2 & u_3 \\ p_1 & p_2 & p_3 \\ c_1 & c_2 & c_3 \end{vmatrix} + \begin{vmatrix} u_1 & u_2 & u_3 \\ q_1 & q_2 & q_3 \\ c_1 & c_2 & c_3 \end{vmatrix} + \begin{vmatrix} v_1 & v_2 & v_3 \\ p_1 & p_2 & p_3 \\ c_1 & c_2 & c_3 \end{vmatrix} + \ldots.$$

(vi) *The value of the determinant is unaltered if to any one row (column) are added multiples of the other rows (columns).* That is

$$\begin{vmatrix} a_1+pa_2+qa_3 & b_1+pb_2+qb_3 & c_1+pc_2+qc_3 \\ a_2 & b_2 & c_2 \\ a_3 & b_3 & c_3 \end{vmatrix} = \begin{vmatrix} a_1 & b_1 & c_1 \\ a_2 & b_2 & c_2 \\ a_3 & b_3 & c_3 \end{vmatrix}.$$

The left-hand side, by (v) and (iv), is

$$\begin{vmatrix} a_1 & b_1 & c_1 \\ a_2 & b_2 & c_2 \\ a_3 & b_3 & c_3 \end{vmatrix} + p \begin{vmatrix} a_2 & b_2 & c_2 \\ a_2 & b_2 & c_2 \\ a_3 & b_3 & c_3 \end{vmatrix} + q \begin{vmatrix} a_3 & b_3 & c_3 \\ a_2 & b_2 & c_2 \\ a_3 & b_3 & c_3 \end{vmatrix},$$

and this, by (iii), is

$$\begin{vmatrix} a_1 & b_1 & c_1 \\ a_2 & b_2 & c_2 \\ a_3 & b_3 & c_3 \end{vmatrix}.$$

Examples 2

1. Use properties (v), (ii), (iii) to prove that the determinant

$$\begin{vmatrix} a_1+\lambda b_1 & b_1+\mu c_1 & c_1+\nu a_1 \\ a_2+\lambda b_2 & b_2+\mu c_2 & c_2+\nu a_2 \\ a_3+\lambda b_3 & b_3+\mu c_3 & c_3+\nu a_3 \end{vmatrix}$$

vanishes whenever the product $\lambda\mu\nu$ has a certain value, to be determined.

2. Prove by property (v) that the determinant

$$\begin{vmatrix} 1+x & 2+x & 3+x \\ 2+x & 3+x & 4+x \\ 3+x & 5+x & x \end{vmatrix}$$

is of the form $A+Bx$, where A, B are certain constants, to be determined.

6. The evaluation of numerical determinants with large numbers. The results of the preceding paragraph may be used to evaluate determinants for which calculation straight from the definition would prove awkward. To illustrate the method, the calculations are given first and comments afterwards.

(i) Let

$$\Delta \equiv \begin{vmatrix} 17 & 36 & 23 \\ 8 & 17 & 10 \\ 24 & 55 & 31 \end{vmatrix}.$$

Subtract 2 times the second row from the first. Then (§5, (vi))

$$\Delta = \begin{vmatrix} 1 & 2 & 3 \\ 8 & 17 & 10 \\ 24 & 55 & 31 \end{vmatrix}.$$

Subtract 3 times the second row from the third. Then

$$\Delta = \begin{vmatrix} 1 & 2 & 3 \\ 8 & 17 & 10 \\ 0 & 4 & 1 \end{vmatrix}.$$

Subtract 8 times the first row from the second. Then

$$\Delta = \begin{vmatrix} 1 & 2 & 3 \\ 0 & 1 & -14 \\ 0 & 4 & 1 \end{vmatrix}.$$

Expand in terms of the first column. Then (§5, (i))

$$\begin{aligned} \Delta &= 1\begin{vmatrix} 1 & -14 \\ 4 & 1 \end{vmatrix} - 0 + 0 \\ &= 1+56 \\ &= 57. \end{aligned}$$

COMMENTS. The first and second steps of the calculation have the obvious aim of reducing numbers. The third step, though similar in result, is intended primarily to obtain the column

$$\begin{matrix} 1 \\ 0 \\ 0 \end{matrix}$$

from which the last step (the goal of the whole calculation) is immediate.

Note the brief descriptions of the steps; it is most important to insert some such indication of what is being done, as the reader (or, possibly, examiner) may otherwise be unable to follow the reasoning. (Sometimes the notation r_1, r_2, r_3 for the rows and c_1, c_2, c_3 for the columns is used; the first three steps are then r_1-2r_2; r_3-3r_2; r_2-8r_1.)

(ii) Consider next the determinant

$$\Delta = \begin{vmatrix} 14 & 29 & 45 \\ 5 & -2 & 19 \\ 9 & 17 & 25 \end{vmatrix}.$$

Subtract 2 times the first column from the second; then 3 times the first column from the third:

$$\Delta = \begin{vmatrix} 14 & 1 & 3 \\ 5 & -12 & 4 \\ 9 & -1 & -2 \end{vmatrix}.$$

From the first row take the sum of the second and third:

$$\Delta = \begin{vmatrix} 0 & 14 & 1 \\ 5 & -12 & 4 \\ 9 & -1 & -2 \end{vmatrix}.$$

Subtract 14 times the third column from the second:

$$\Delta = \begin{vmatrix} 0 & 0 & 1 \\ 5 & -68 & 4 \\ 9 & 27 & -2 \end{vmatrix}.$$

Expand in terms of the first row:

$$\Delta = 0-0+\begin{vmatrix} 5 & -68 \\ 9 & 27 \end{vmatrix}$$

$$= 135+612$$

$$= 747.$$

IMPORTANT NOTE. With practice, several steps may be performed simultaneously, but it is advisable to arrange the work so that at each stage *at least one row or one column is kept unaltered.* The following general example shows the danger of combining too many steps:

Let

$$\Delta \equiv \begin{vmatrix} a_1 & b_1 & c_1 \\ a_2 & b_2 & c_2 \\ a_3 & b_3 & c_3 \end{vmatrix}.$$

Subtract the second column from the first, the third from the second, the first from the third:

$$\Delta = \begin{vmatrix} a_1-b_1 & b_1-c_1 & c_1-a_1 \\ a_2-b_2 & b_2-c_2 & c_2-a_2 \\ a_3-b_3 & b_3-c_3 & c_3-a_3 \end{vmatrix}.$$

Add the second and third columns to the first:

$$\Delta = \begin{vmatrix} 0 & b_1-c_1 & c_1-a_1 \\ 0 & b_2-c_2 & c_2-a_2 \\ 0 & b_3-c_3 & c_3-a_3 \end{vmatrix}.$$

Expand in terms of the first column:

$$\Delta = 0.$$

Hence, apparently, the value of any determinant is zero.

The error would have been avoided by using as the start:

Subtract the second column from the first, the third from the second, *leaving the third unaltered.*

Examples 3

1. Evaluate the determinants of three rows and columns on p. 247 by reducing each of them to a form in which a row or a column has two zeros.

Evaluate the determinants

2. $\begin{vmatrix} 24 & 25 & 26 \\ 27 & 29 & 31 \\ 30 & 33 & 36 \end{vmatrix}$.

3. $\begin{vmatrix} 14 & -10 & 6 \\ -17 & 15 & -13 \\ 20 & -18 & 16 \end{vmatrix}$.

4. $\begin{vmatrix} 5 & 26 & 46 \\ 7 & 38 & 26 \\ 4 & 19 & 32 \end{vmatrix}$.

5. $\begin{vmatrix} 4 & 8 & 6 \\ 9 & 5 & -3 \\ -2 & 7 & 4 \end{vmatrix}$.

6. $\begin{vmatrix} -3 & 7 & 4 \\ 2 & -8 & 6 \\ 9 & 5 & 1 \end{vmatrix}$.

7. $\begin{vmatrix} 3 & 0 & 17 \\ 15 & 13 & -9 \\ 2 & -15 & 4 \end{vmatrix}$.

8. $\begin{vmatrix} 3 & 19 & 22 \\ 5 & 31 & 37 \\ 2 & 14 & 13 \end{vmatrix}$.

9. $\begin{vmatrix} 5 & -4 & 3 \\ -2 & 1 & 7 \\ 9 & -11 & 15 \end{vmatrix}$.

7. Cofactors. The greek letter Δ is often used to name a typical determinant; thus

$$\Delta \equiv \begin{vmatrix} a_1 & b_1 & c_1 \\ a_2 & b_2 & c_2 \\ a_3 & b_3 & c_3 \end{vmatrix}.$$

Nine determinants, each of two rows and columns, can be formed from Δ by selecting any one of its elements and then deleting the row and column through it. Such a determinant is called the *minor* of the element selected. Instead of minors, however, it is found better to use *cofactors*, which are equal to the minors numerically, but which carry the sign + or − according to the place occupied by the selected element in the scheme

$$\begin{vmatrix} + & - & + \\ - & + & - \\ + & - & + \end{vmatrix}.$$

The cofactor is named by using the capital letter of the corresponding element.

For example,

$$A_1 \equiv \begin{vmatrix} b_2 & c_2 \\ b_3 & c_3 \end{vmatrix}, \quad C_2 \equiv -\begin{vmatrix} a_1 & b_1 \\ a_3 & b_3 \end{vmatrix}, \quad B_3 \equiv -\begin{vmatrix} a_1 & c_1 \\ a_2 & c_2 \end{vmatrix}.$$

Note that *the cofactor of an element is its coefficient in the usual expansion of the determinant.*

8. Expansion by cofactors; row-substitution and column-substitution. Let the cofactors of the elements of any one row (column) of Δ be selected, and let u, v, w be any three numbers. To prove that, *if the cofactors, in their order in the row (column) are multiplied by u, v, w respectively and added, then the result is of the same form as Δ, but with u, v, w replacing the elements of the selected row (column).*

For example, select the elements a_2, b_2, c_2, with cofactors A_2, B_2, C_2. We have to prove the formula

$$uA_2+vB_2+wC_2 = \begin{vmatrix} a_1 & b_1 & c_1 \\ u & v & w \\ a_3 & b_3 & c_3 \end{vmatrix}.$$

The determinant on the right is (p. 246)

$$\begin{aligned} a_1vc_3-a_1wb_3+b_1wa_3-b_1uc_3+c_1ub_3-c_1va_3 \\ &= -u(b_1c_3-b_3c_1)+v(a_1c_3-a_3c_1)-w(a_1b_3-a_3b_1) \\ &= uA_2+vB_2+wC_2 \\ &= \text{the left-hand side.} \end{aligned}$$

(This is merely elaborating the intuitive argument that u, v, w stand exactly in place of a_2, b_2, c_2 without affecting the rest of the determinant.)

We call this result *the law of row-substitution (column-substitution)*, since its effect is to substitute the numbers u, v, w for the row (column) selected.

COROLLARIES: (i) If u, v, w are the actual elements of the row (column) selected, the right-hand side is the given determinant Δ. Hence there are *six identities*

$$\begin{aligned} a_1A_1+b_1B_1+c_1C_1 &= \Delta, \quad & a_1A_1+a_2A_2+a_3A_3 &= \Delta, \\ a_2A_2+b_2B_2+c_2C_2 &= \Delta, \quad & b_1B_1+b_2B_2+b_3B_3 &= \Delta, \\ a_3A_3+b_3B_3+c_3C_3 &= \Delta, \quad & c_1C_1+c_2C_2+c_3C_3 &= \Delta. \end{aligned}$$

(ii) If u, v, w are the elements of any row (column) *other than the one selected*, then the right-hand side is a determinant with two equal rows (columns), so that its value is zero. Hence there are *twelve identities*, of which the following four are typical:

$$a_1A_2+b_1B_2+c_1C_2 = 0, \qquad a_1B_1+a_2B_2+a_3B_3 = 0,$$
$$a_1A_3+b_1B_3+c_1C_3 = 0, \qquad a_1C_1+a_2C_2+a_3C_3 = 0.$$

9. Determinants of higher order.* Determinants of order higher than three may be defined successively by an inductive process. We confine ourselves to four rows and columns. The determinant

$$\Delta \equiv \begin{vmatrix} a_1 & b_1 & c_1 & d_1 \\ a_2 & b_2 & c_2 & d_2 \\ a_3 & b_3 & c_3 & d_3 \\ a_4 & b_4 & c_4 & d_4 \end{vmatrix}$$

is defined as follows:

Let $A_1, -B_1, C_1, -D_1$, with alternating signs, be the determinants of three rows and columns formed by omitting the first row and the column with a_1, b_1, c_1, d_1 respectively. Then Δ is defined by the identity

$$\Delta \equiv a_1A_1+b_1B_1+c_1C_1+d_1D_1.$$

The determinants A_1, B_1, C_1, D_1 are called the cofactors of a_1, b_1, c_1, d_1 in Δ. More generally, the cofactors $A_1, B_1, \ldots, C_4, D_4$ of the elements $a_1, b_1, \ldots, c_4, d_4$ are defined, for any one of them, to be the determinant obtained by omitting the row and the column containing the element and giving to that determinant the sign + or − according to the scheme

$$\begin{vmatrix} + & - & + & - \\ - & + & - & + \\ + & - & + & - \\ - & + & - & + \end{vmatrix}$$

for the positions

$$\begin{vmatrix} a_1 & b_1 & c_1 & d_1 \\ a_2 & b_2 & c_2 & d_2 \\ a_3 & b_3 & c_3 & d_3 \\ a_4 & b_4 & c_4 & d_4 \end{vmatrix}.$$

The rules given earlier for the manipulation of determinants of three rows and columns are applicable to four. A number of examples are given later in the chapter, but they can be omitted if this paragraph is postponed.

* To be postponed, if desired.

10. The solution of three simultaneous equations. Let

$$\begin{aligned} a_1x+b_1y+c_1z &= d_1, \\ a_2x+b_2y+c_2z &= d_2, \\ a_3x+b_3y+c_3z &= d_3 \end{aligned}$$

be three given equations. Write

$$\Delta \equiv \begin{vmatrix} a_1 & b_1 & c_1 \\ a_2 & b_2 & c_2 \\ a_3 & b_3 & c_3 \end{vmatrix},$$

and name cofactors as in §7.

To solve these equations, multiply them by A_1, A_2, A_3 and add, using the law of column-substitution and the corollaries given in §8. Then

$$\begin{aligned} \Delta x &= d_1A_1+d_2A_2+d_3A_3 \\ &= \begin{vmatrix} d_1 & b_1 & c_1 \\ d_2 & b_2 & c_2 \\ d_3 & b_3 & c_3 \end{vmatrix}. \end{aligned}$$

Similarly

$$\Delta y = \begin{vmatrix} a_1 & d_1 & c_1 \\ a_2 & d_2 & c_2 \\ a_3 & d_3 & c_3 \end{vmatrix},$$

$$\Delta z = \begin{vmatrix} a_1 & b_1 & d_1 \\ a_2 & b_2 & d_2 \\ a_3 & b_3 & d_3 \end{vmatrix}.$$

If $\Delta \neq 0$, these three relations give x, y, z, so that the equations are solved.

NOTE. When the coefficients are purely numerical, this solution tends to be cumbersome, with four determinants to evaluate. When some or all of the coefficients are algebraic, it can be very powerful.

Illustration* 5. *Non-homogeneous equations with vanishing determinant. The evaluation of the determinant of the coefficients reinforces powerfully the method of solution given earlier (p. 88) for three simultaneous equations in the exceptional cases. Consider, for example, the equations

$$\begin{aligned} x+y+z &= 3, \\ 2x+3y+4z &= 9, \\ ax+2y+z &= b. \end{aligned}$$

The determinant of the coefficients is

$$\Delta \equiv \begin{vmatrix} 1 & 1 & 1 \\ 2 & 3 & 4 \\ a & 2 & 1 \end{vmatrix} = a-3.$$

The solution is therefore straight-forward except when $a = 3$.

When $a = 3$ the set of equations is

$$x+y+z = 3,$$
$$2x+3y+4z = 9,$$
$$3x+2y+z = b.$$

To solve, eliminate x between two pairs of equations. From the second subtract twice the first:

$$y+2z = 3;$$

from the third subtract three times the first:

$$-y-2z = b-9.$$

This gives two values for $y+2z$, namely 3 and $9-b$, and so *the equations cannot be solved unless these values are equal*, that is, unless

$$b = 6.$$

Hence *there is no solution when* $a = 3$, $b \neq 6$.

When $b = 6$, elimination of x as above gives only the one relation

$$y+2z = 3$$

connecting y and z. They cannot therefore be found uniquely.

If z is given an arbitrary value λ, then

$$y = 3-2\lambda,$$

and so (since $x+y+z = 3$), $\qquad x = \lambda.$

Thus *when* $a = 3$, $b = 6$, *there is an infinity of solutions*, in the form

$$x = \lambda, \quad y = 3-2\lambda, \quad z = \lambda,$$

where λ is any arbitrary number.

Note that *the condition* $\Delta = 0$ *has served to isolate the difficult cases.* At the present level of his work, the student will probably find it best to adopt the procedure:

(i) Calculate Δ and equate it to zero to find the conditions of exception;

(ii) Return to the given equations, solving by elimination as in the earlier parts of this book, but using the conditions obtained in (i).

It may be remarked that, in the cases which usually arise, the determinant Δ can be factorized. This seems to lead to more interesting sets of equations.

Examples 4

(The following examples, in which the arithmetic is very simple, are inserted to enable the reader to become familiar with the underlying ideas. They can all be solved easily without determinants if desired.)

1. Prove that the equations

$$x+y = 3,$$
$$3x-y = 1,$$
$$x-y = 0$$

are not consistent.

2. Prove that the equations

$$2x+3y = 5,$$
$$x+y = 2,$$
$$3x-y = k$$

are consistent provided that k has a certain value, to be found.

3. Determine whether there is a value of k for which the equations

$$2x+y = 1,$$
$$3x+y = 2,$$
$$x+ky = 1$$

are consistent.

4. Prove that the equations

$$x+y = 2,$$
$$4x-3y = 1,$$
$$2x+ay = b$$

can be satisfied simultaneously if, and only if, $a-b+2 = 0$.

5. Prove that the equations

$$x-y = 2,$$
$$2x+ay = b$$

can be solved uniquely for x and y except when $a = -2$.

Prove that, when $a = -2$, they cannot be solved at all except when also $b = 4$.

Prove that, when $a = -2$, $b = 4$, the equations are satisfied when $x = \lambda$, $y = \lambda-2$ for all values of λ.

6. Examine, after the manner of Question 5, the equations

$$\text{(i)}\quad x+y = 3,\quad ax+y = b. \qquad \text{(ii)}\quad 2x+y = 5,\quad x+ay = b. \qquad \text{(iii)}\quad x-3y = 4,\quad ax+2y = b.$$

7. Find the values of k for which the equations

$$x+3y = k,$$
$$x+ky = 3,$$
$$kx+9y = 9$$

are consistent.

8. Examine for consistency the equations

$$2x-y = k,$$
$$2x-ky = 1,$$
$$2kx-y = 1.$$

11. Three homogeneous simultaneous equations. Let

$$a_1x+b_1y+c_1z = 0,$$
$$a_2x+b_2y+c_2z = 0,$$
$$a_3x+b_3y+c_3z = 0$$

be three equations each homogeneous in x, y, z and denote by Δ the determinant

$$\Delta \equiv \begin{vmatrix} a_1 & b_1 & c_1 \\ a_2 & b_2 & c_2 \\ a_3 & b_3 & c_3 \end{vmatrix}.$$

The solution $x = y = z = 0$ is trivial, and the aim now is to discuss solutions in which x, y, z are not all zero.

(i) *To prove that, if there are solutions in which x, y, z are not all zero, then $\Delta = 0$.*

Multiply the equations by A_1, A_2, A_3 and add. Then (p. 256)

$$\Delta x = 0.$$

Similarly, $$\Delta y = 0, \quad \Delta z = 0.$$

Since x, y, z are not all zero, it follows that

$$\Delta = 0.$$

NOTE: This result means that Δ is the *eliminant* of the given equations: see p. 245.

(ii) To prove that, *if* $\Delta = 0$, *then there are solutions in which* x, y, z *are not all zero.*

(*a*) Suppose first that *not all cofactors are zero*; say, for precision, $A_1 \neq 0$.

The two equations

$$a_2x+b_2y+c_2z = 0,$$
$$a_3x+b_3y+c_3z = 0$$

certainly have a solution; for example (by p. 255),

$$x = A_1, \quad y = B_1, \quad z = C_1,$$

and the value of x is not then zero. But this solution also satisfies the first equation, since (p. 254)

$$a_1A_1+b_1B_1+c_1C_1 = \Delta$$
$$= 0,$$

given. The three equations are therefore satisfied by

$$x = kA_1, \quad y = kB_1, \quad z = kC_1$$

for all values of k.

Note also that direct solution of the two equations selected above shows that any solution is necessarily of this form (cf. p. 242).

(*b*)* Suppose next that *all cofactors are zero.* It is implicit that the nine given coefficients are not all zero; say, in particular, $a_1 \neq 0$. Then, using the cofactors C_3 and B_3, we have

$$C_3 = 0, \quad \text{or} \quad a_1b_2-a_2b_1 = 0,$$
$$B_3 = 0, \quad \text{or} \quad a_1c_2-a_2c_1 = 0.$$

If $a_2 = 0$, then (since $a_1 \neq 0$) $b_2 = c_2 = 0$ also. Neglecting this very trivial case, we have

$$\frac{a_1}{a_2} = \frac{b_1}{b_2} = \frac{c_1}{c_2}.$$

Similarly,

$$\frac{a_1}{a_3} = \frac{b_1}{b_3} = \frac{c_1}{c_3}.$$

The three equations thus have their coefficients proportional, so that *there is effectively only one independent equation*, say

$$a_1x+b_1y+c_1z = 0.$$

This has an infinity of solutions, of which an obvious one is

$$x = 0, \quad y = c_1, \quad z = -b_1.$$

The equations are therefore soluble.

* This complication may be omitted at a first reading.

Illustration 6. *To find values of* ρ *such that*

$$\begin{aligned} 11x-6y+2z &= \rho x, \\ -6x+10y-4z &= \rho y, \\ 2x-4y+6z &= \rho z, \end{aligned}$$

and to solve the equations.

Write the equations in the form

$$\begin{aligned} (11-\rho)x \quad -6y \quad +2z &= 0, \\ -6x+(10-\rho)y \quad -4z &= 0, \\ 2x \quad -4y+(6-\rho)z &= 0. \end{aligned}$$

Eliminate (p. 259) the ratios $x:y:z$ from these equations:

$$\begin{vmatrix} 11-\rho & -6 & 2 \\ -6 & 10-\rho & -4 \\ 2 & -4 & 6-\rho \end{vmatrix} = 0.$$

Hence, on expanding,

$$324-180\rho+27\rho^2-\rho^3 = 0,$$

so that

$$\rho = 3, 6, 18.$$

When $\rho = 3$, the first two equations are

$$\begin{aligned} 8x-6y+2z &= 0, \\ -6x+7y-4z &= 0. \end{aligned}$$

so that

$$\frac{x}{1} = \frac{y}{2} = \frac{z}{2}.$$

When $\rho = 6$, the first two equations are

$$\begin{aligned} 5x-6y+2z &= 0, \\ -6x+4y-4z &= 0 \end{aligned}$$

so that

$$\frac{x}{2} = \frac{y}{1} = \frac{z}{-2}.$$

When $\rho = 18$, the first two equations are

$$\begin{aligned} -7x-6y+2z &= 0, \\ -6x-8y-4z &= 0, \end{aligned}$$

so that

$$\frac{x}{2} = \frac{y}{-2} = \frac{z}{1}.$$

Illustrations from geometry. (i) *To find the condition for the three lines*

$$\begin{aligned} l_1x+m_1y+n_1 &= 0, \\ l_2x+m_2y+n_2 &= 0, \\ l_3x+m_3y+n_3 &= 0 \end{aligned}$$

to be concurrent.

If there is a common point, let it be (α, β). Then

$$l_1\alpha+m_1\beta+n_1 = 0,$$
$$l_2\alpha+m_2\beta+n_2 = 0,$$
$$l_3\alpha+m_3\beta+n_3 = 0.$$

Eliminate the ratios $\alpha:\beta:1$ from these equations; thus the condition is

$$\begin{vmatrix} l_1 & m_1 & n_1 \\ l_2 & m_2 & n_2 \\ l_3 & m_3 & n_3 \end{vmatrix} = 0.$$

(ii) *To find the equation of the circle through the three points* (x_1, y_1), (x_2, y_2), (x_3, y_3).

The equation of any circle is

$$x^2+y^2+2gx+2fy+c = 0.$$

For the three points to lie on it, the following conditions must be satisfied:

$$x_1^2+y_1^2+2gx_1+2fy_1+c = 0,$$
$$x_2^2+y_2^2+2gx_2+2fy_2+c = 0,$$
$$x_3^2+y_3^2+2gx_3+2fy_3+c = 0.$$

Eliminate $1:2g:2f:c$ between these four equations. This gives a relation connecting x, y, which is the required equation of the circle, in the form

$$\begin{vmatrix} x^2+y^2 & x & y & 1 \\ x_1^2+y_1^2 & x_1 & y_1 & 1 \\ x_2^2+y_2^2 & x_2 & y_2 & 1 \\ x_3^2+y_3^2 & x_3 & y_3 & 1 \end{vmatrix} = 0.$$

Examples 5

Solve the following sets of examples, giving particular attention to the exceptional cases:

1.
$$x+y+z = 0,$$
$$x+2y+3z = 0,$$
$$2x+y+2z = b.$$

2.
$$ax+y+2z = b,$$
$$x+2y+3z = 0,$$
$$2x+3y+4z = 0.$$

3.
$$3ax-6y+18z = b,$$
$$4x+5y-z = -2,$$
$$x-7y-z = 3.$$

4.
$$x+y+z = 3,$$
$$x-y+2z = 4,$$
$$ax+4y+z = 5.$$

5.
$$ax+4y+2z = b,$$
$$4x+7y+3z = 8,$$
$$2x+y+z = 2.$$

6.
$$x+y+az = b+3,$$
$$x-3y+2z = -1,$$
$$2x+6y+z = 4.$$

7.
$$(1+a)(y+z) = 2,$$
$$x+ay+z = 2,$$
$$x+y+az = 2.$$

8.
$$ax+y+z = p,$$
$$x+ay+z = q,$$
$$x+y+az = r.$$

9.
$$ax+3y+2z = b,$$
$$5x+4y+3z = 1,$$
$$x+2y+z = b^2.$$

10.
$$x+y+kz = 4k,$$
$$x+ky+z = -2,$$
$$2x+y+z = -2.$$

11.
$$(b+c)x+a(y+z) = a,$$
$$(c+a)y+b(z+x) = b,$$
$$(a+b)z+c(x+y) = c.$$

12.
$$ax+by+cz = 1,$$
$$bx+cy+az = 1,$$
$$cx+ay+bz = 1,$$

where a, b, c are real.

13.
$$(1+a)x+by+cz = 1,$$
$$ax+(1+b)y+cz = 0,$$
$$ax+by+(1+c)z = 0.$$

Show also that, in the case when $a+b+c=0$, the solutions can be expressed in a form depending on a only.

14.
$$x+y+z = 1,$$
$$ax+by+cz = 2,$$
$$a^3x+b^3y+c^3z = 8,$$
where a, b, c are unequal.

15.
$$a^2x+ay+z = a^2,$$
$$ax+y+bz = 1,$$
$$a^2bx+y+bz = b.$$

16. Find the values of a for which the equations
$$ax+y+4z = 6,$$
$$x+ay+z = 3,$$
$$2x+2y+az = 5,$$
$$x+y-z = 1$$
are consistent, and give the solution in each case.

17. Find all the values of the constant λ for which the equations
$$x+2y+z = \lambda x,$$
$$2x+y+z = \lambda y,$$
$$x+y+2z = \lambda z$$
can be satisfied by values of x, y, z not all zero.
Determine the ratios $x:y:z$ for the values of λ.

18. Find the values of λ such that the equations
$$2x \quad -4z = \lambda x,$$
$$2y-4z = \lambda y,$$
$$-4x-4y-2z = \lambda z$$
have a solution other than $x = y = z = 0$, and find the corresponding ratios $x:y:z$.

19. Show that, if no two of a, b, c are equal, and if
$$bc+ca+ab = 0,$$
then the complete solution of the equations
$$x+y+z = 3,$$
$$a^2x+b^2y+c^2z = a^2+b^2+c^2,$$
$$a^3x+b^3y+c^3z = a^3+b^3+c^3$$
is given by
$$x = 1+\lambda(b^2-c^2), \quad y = 1+\lambda(c^2-a^2), \quad z = 1+\lambda(a^2-b^2).$$

20. In the equations

$$\begin{aligned} -ny + mz &= a, \\ nx \quad - lz &= b, \\ -mx + ly \quad &= c, \\ lx + my + nz &= p, \end{aligned}$$

l, m, n, a, b, c, p are given real numbers (l, m, n not all zero). Prove that a *necessary* and *sufficient* condition for the equations to have a solution is

$$la + mb + nc = 0,$$

and solve the equations when this condition is satisfied.

12. Determinants as polynomials; factors. It is sometimes possible to evaluate determinants having a recognizable pattern without going through all the steps of a detailed calculation. The remainder theorem (p. 44) is often useful in this connection.

Consider, for example, the determinantal equation

$$\begin{vmatrix} x^2 & x & 1 \\ 4 & 2 & 1 \\ 9 & -3 & 1 \end{vmatrix} = 0.$$

It is at once obvious that the left-hand side is zero when $x = 2$, since then the first and second rows are equal; also that it is zero when $x = -3$, since then the first and third rows are equal. Moreover, the equation is quadratic in x, on expansion in terms of the first row, and so there are only two roots to be found. Hence the solution is

$$x = 2, \quad x = -3.$$

In precisely the same way, the solution of the cubic equation

$$\begin{vmatrix} x^3 & x^2 & x & 1 \\ -1 & 1 & -1 & 1 \\ 8 & 4 & 2 & 1 \\ 27 & 9 & 3 & 1 \end{vmatrix} = 0$$

is

$$x = -1, \quad x = 2, \quad x = 3.$$

Consider, further, the equation

$$\begin{vmatrix} x^3 & x & 1 \\ 8 & 2 & 1 \\ 125 & 5 & 1 \end{vmatrix} = 0.$$

It is satisfied, as above, for $x = 2$, $x = 5$. Moreover, by expansion of the first row, it is a cubic of the form

$$Ax^3+Bx+C = 0,$$

so that (p. 60) the sum of the roots is zero. The third root is thus -7, so the solution is

$$x = 2, \quad x = 5, \quad x = -7.$$

Examples 6

Solve mentally the equations:

1. $\begin{vmatrix} x & 1 \\ 5 & 1 \end{vmatrix} = 0.$ **2.** $\begin{vmatrix} x & 1 \\ -3 & 1 \end{vmatrix} = 0.$ **3.** $\begin{vmatrix} x & 1 \\ 5 & 2 \end{vmatrix} = 0.$

4. $\begin{vmatrix} x^2 & x & 1 \\ 1 & -1 & 1 \\ 4 & 2 & 1 \end{vmatrix} = 0.$ **5.** $\begin{vmatrix} x^2 & x & 1 \\ a^2 & a & 1 \\ 1 & 1 & 1 \end{vmatrix} = 0.$

6. $\begin{vmatrix} x^2 & x & 1 \\ a^2 & a & 1 \\ b^2 & b & 1 \end{vmatrix} = 0.$ **7.** $\begin{vmatrix} x^3 & x & 1 \\ 1 & 1 & 1 \\ 8 & 2 & 1 \end{vmatrix} = 0.$

8. $\begin{vmatrix} x^3 & x & 1 \\ a^3 & a & 1 \\ 8 & 2 & 1 \end{vmatrix} = 0.$ **9.** $\begin{vmatrix} x^3 & x & 1 \\ a^3 & a & 1 \\ b^3 & b & 1 \end{vmatrix} = 0.$

10. $\begin{vmatrix} x^3 & x^2 & 1 \\ 8 & 4 & 1 \\ 27 & 9 & 1 \end{vmatrix} = 0.$ **11.** $\begin{vmatrix} x^3 & x^2 & 1 \\ a^3 & a^2 & 1 \\ b^3 & b^2 & 1 \end{vmatrix} = 0.$

***Illustration* 7.** *To factorize the determinant*

$$\Delta \equiv \begin{vmatrix} 1 & a & a^3 \\ 1 & b & b^3 \\ 1 & c & c^3 \end{vmatrix}.$$

Interchange of a, b gives a determinant identical with Δ except that two rows are interchanged; it is thus the same as Δ apart from a change of sign. Moreover Δ is, on expansion, a polynomial, which is skew-symmetric in a, b. It is, similarly, skew-symmetric in a, b, c, and so (p. 59) has

$$(b-c)(c-a)(a-b)$$

as a factor.

We thus have

(i) Δ, a skew-symmetric polynomial of degree 4 on expansion.

(ii) The expression $(b-c)(c-a)(a-b)$, which is a skew-symmetric polynomial of degree 3 and a factor of Δ.

The remaining factor of Δ is therefore *symmetrical* in a, b, c and of degree 1, so that it is necessarily of the form

$$A(a+b+c),$$

where A is numerical. Hence

$$\Delta \equiv A(b-c)(c-a)(a-b)(a+b+c).$$

Now the expansion of Δ contains the term

$$+bc^3,$$

which must consequently appear in the expansion of the right-hand side. To get one b and three c's we need the terms in clarendon type

$$A(b-\mathbf{c})(\mathbf{c}-a)(a-\mathbf{b})(+a+b+\mathbf{c}),$$

so that

$$1 = A(-1)(+1)(-1)(+1),$$

or

$$A = 1.$$

Hence

$$\Delta = (b-c)(c-a)(a-b)(a+b+c).$$

***Illustration* 8.** *To factorize the determinant*

$$\Delta \equiv \begin{vmatrix} a & b & c & d \\ d & a & b & c \\ c & d & a & b \\ b & c & d & a \end{vmatrix}.$$

Let θ be any root of the equation

$$\theta^4 = 1.$$

To the first column add θ times the second, θ^2 times the third and θ^3 times the fourth. Then

$$\Delta \equiv \begin{vmatrix} a+\theta b+\theta^2 c+\theta^3 d & b & c & d \\ d+\theta a+\theta^2 b+\theta^3 c & a & b & c \\ c+\theta d+\theta^2 a+\theta^3 b & d & a & b \\ b+\theta c+\theta^2 d+\theta^3 a & c & d & a \end{vmatrix}.$$

Hence

$$\Delta \equiv \begin{vmatrix} a+\theta b+\theta^2 c+\theta^3 d & b & c & d \\ \theta(a+\theta b+\theta^2 c+\theta^3 d) & a & b & c \\ \theta^2(a+\theta b+\theta^2 c+\theta^3 d) & d & a & b \\ \theta^3(a+\theta b+\theta^2 c+\theta^3 d) & c & d & a \end{vmatrix}$$

in virtue of the relations $1 = \theta^4$, $\theta = \theta^5$, $\theta^2 = \theta^6$. Each element of the first column now has a factor

$$a+\theta b+\theta^2 c+\theta^3 d,$$

and so (p. 248) this is a factor of Δ. Since θ may have any one of the four values $1, -1, i, -i$ ($i^2 = -1$), this gives four factors each linear in a, b, c, d. But Δ is a polynomial of degree 4 in a, b, c, d, so that the remaining factor, if any, is purely numerical, say N. Hence there is an identity

$$\Delta \equiv N(a+b+c+d)(a-b+c-d)(a+ib-c-id)(a-ib-c+id).$$

Since N is independent of a, b, c, d, we may consider the special case $a = 1, b = c = d = 0$; this gives

$$N = 1.$$

Hence, in real form,

$$\Delta \equiv (a+b+c+d)(a-b+c-d)\{(a-c)^2+(b-d)^2\}.$$

Examples 7

1. Prove that, if

$$\Delta = \begin{vmatrix} a & b & c \\ c & a & b \\ b & c & a \end{vmatrix},$$

then (i) $\Delta \equiv a^3+b^3+c^3-3abc$,

(ii) $\Delta \equiv (a+b+c)(a+\omega b+\omega^2 c)(a+\omega^2 b+\omega c)$,

where $\omega^3 = 1, \omega \neq 1$.

2. Factorize the determinants:

(i) $\begin{vmatrix} 1 & 1 & 1 \\ a & b & c \\ a^2 & b^2 & c^2 \end{vmatrix}$, (ii) $\begin{vmatrix} 1 & 1 & 1 \\ a & b & c \\ bc & ca & ab \end{vmatrix}$,

(iii) $\begin{vmatrix} a & b & c \\ a^2 & b^2 & c^2 \\ a^3 & b^3 & c^3 \end{vmatrix}$, (iv) $\begin{vmatrix} 1 & 1 & 1 \\ bc & ca & ab \\ a^2 & b^2 & c^2 \end{vmatrix}$.

13. The multiplication of determinants.

There is a basic difficulty about finding an expression for the product of two determinants. The rules of §5 (p. 248) enable us to express a given determinant in many different ways, for example, by changing the orders of rows or columns, or by interchanging rows and columns. The final answer for the product is definite, but it may appear in different forms as a result of these preliminary manipulations, and the proof therefore consists (at this stage) of verifying a proposed form rather than of proceeding directly to a form which grows of itself out of the process of multiplication. The present treatment is based on 'matrix multiplication', which will become important later for the student who continues with linear algebra.

Let

$$\Delta \equiv \begin{vmatrix} a_1 & b_1 & c_1 \\ a_2 & b_2 & c_2 \\ a_3 & b_3 & c_3 \end{vmatrix}, \quad \Delta' \equiv \begin{vmatrix} p_1 & q_1 & r_1 \\ p_2 & q_2 & r_2 \\ p_3 & q_3 & r_3 \end{vmatrix}$$

be two given determinants, whose product is required. In order to work with a convenient notation, interchange rows and columns in Δ', thus expressing it in the form

$$\Delta' \equiv \begin{vmatrix} p_1 & p_2 & p_3 \\ q_1 & q_2 & q_3 \\ r_1 & r_2 & r_3 \end{vmatrix}.$$

For brevity in writing, use the notation

$$(ij)$$

to denote the expression $a_i p_j + b_i q_j + c_i r_j$,

where the suffix i goes with a, b, c and the suffix j with p, q, r. For example,

$$(11) \equiv a_1 p_1 + b_1 q_1 + c_1 r_1, \quad (32) \equiv a_3 p_2 + b_3 q_2 + c_3 r_2.$$

Then *the formula for the product is*

$$\Delta\Delta' = \begin{vmatrix} (11) & (12) & (13) \\ (21) & (22) & (23) \\ (31) & (32) & (33) \end{vmatrix}.$$

In full, the right-hand side is

$$\begin{vmatrix} a_1p_1+b_1q_1+c_1r_1 & a_1p_2+b_1q_2+c_1r_2 & a_1p_3+b_1q_3+c_1r_3 \\ a_2p_1+b_2q_1+c_2r_1 & a_2p_2+b_2q_2+c_2r_2 & a_2p_3+b_2q_3+c_2r_3 \\ a_3p_1+b_3q_1+c_3r_1 & a_3p_2+b_3q_2+c_3r_2 & a_3p_3+b_3q_3+c_3r_3 \end{vmatrix}.$$

The elements of the columns are expressed in composite form, and so (p. 249; compare the Corollary) this determinant can be expressed as the sum of 27 determinants. Among the 27 there are 21 of the type

$$\begin{vmatrix} a_1p_1 & a_1p_2 & b_1q_3 \\ a_2p_1 & a_2p_2 & b_2q_3 \\ a_3p_1 & a_3p_2 & b_3q_3 \end{vmatrix},$$

where two or more of the columns have the same letter-pairs a, p; b, q; or c, r though with differing suffixes. Such determinants are zero, since, for example, the one quoted (after taking factors p_1, p_2, q_3 from the first, second and third columns respectively) is

$$p_1p_2q_3 \begin{vmatrix} a_1 & a_1 & b_1 \\ a_2 & a_2 & b_2 \\ a_3 & a_3 & b_3 \end{vmatrix},$$

having two equal columns.

The remaining six determinants are (quoting the top rows only, for brevity)

$$|\,a_1p_1 \quad b_1q_2 \quad c_1r_3\,|, \quad |\,a_1p_1 \quad c_1r_2 \quad b_1q_3\,|, \quad |\,b_1q_1 \quad a_1p_2 \quad c_1r_3\,|,$$

$$|\,b_1q_1 \quad c_1r_2 \quad a_1p_3\,|, \quad |\,c_1r_1 \quad a_1p_2 \quad b_1q_3\,|, \quad |\,c_1r_1 \quad b_1q_2 \quad a_1p_3\,|,$$

or, after taking out factors as before and then re-arranging the orders of columns where necessary,

$$p_1q_2r_3\Delta,\ -p_1r_2q_3\Delta,\ -q_1p_2r_3\Delta,$$
$$q_1r_2p_3\Delta,\quad r_1p_2q_3\Delta,\ -r_1q_2p_3\Delta.$$

Adding these six expressions, we obtain the right-hand side in the form of the product

$$\Delta\Delta'.$$

NOTE. The corresponding rule for determinants with two rows and columns is

$$\begin{vmatrix} a_1 & b_1 \\ a_2 & b_2 \end{vmatrix} \begin{vmatrix} p_1 & p_2 \\ q_1 & q_2 \end{vmatrix} = \begin{vmatrix} a_1p_1+b_1q_1 & a_1p_2+b_1q_2 \\ a_2p_1+b_2q_1 & a_2p_2+b_2q_2 \end{vmatrix}.$$

Remark on a dynamic method of multiplication.

What we have proved, essentially, is the formula

$$\begin{vmatrix} a_1 & b_1 & c_1 \\ a_2 & b_2 & c_2 \\ a_3 & b_3 & c_3 \end{vmatrix} \begin{vmatrix} p_1 & p_2 & p_3 \\ q_1 & q_2 & q_3 \\ r_1 & r_2 & r_3 \end{vmatrix} = \begin{vmatrix} (11) & (12) & (13) \\ (21) & (22) & (23) \\ (31) & (32) & (33) \end{vmatrix}$$

in which, so to speak, the rows of the first are combined with the columns of the second. To get the element in, say, the second row and the third column of the product—that is, the term (23)—we go *along the second row and down the third column* in the manner implied by the scheme:

$$\begin{vmatrix} \cdot & \cdot & \cdot \\ & \longrightarrow & \\ \cdot & \cdot & \cdot \end{vmatrix} \begin{vmatrix} \cdot & \cdot & \\ \cdot & \cdot & \downarrow \\ \cdot & \cdot & \end{vmatrix}.$$

If the left hand moves along the horizontal arrow while the right hand descends the vertical arrow at the same speed, the three terms of the element (23) are obtained.

Examples

There is no real advantage to be gained in multiplying by this rule two determinants whose elements are all numbers, since it is usually easier to evaluate each directly and then to multiply the answers. For practice, however, the reader should take pairs of determinants from those evaluated on p. 253, multiply by the rule, and then check that the product agrees.

A set of further examples will be found on p. 273 after the following Illustrations.

Illustration 9. *To prove the identity*

$$(a^2+b^2)(c^2+d^2) = (ac-bd)^2+(bc+ad)^2.$$

The factors on the left can be expressed in the form

$$a^2+b^2 = \begin{vmatrix} a & ib \\ ib & a \end{vmatrix},$$

$$c^2+d^2 = \begin{vmatrix} c & id \\ id & c \end{vmatrix}.$$

Hence
$$(a^2+b^2)(c^2+d^2) = \begin{vmatrix} ac+(ib)(id) & a(id)+(ib)c \\ (ib)c+a(id) & (ib)(id)+ac \end{vmatrix}$$

$$= \begin{vmatrix} ac-bd & i(bc+ad) \\ i(bc+ad) & ac-bd \end{vmatrix}$$

$$= (ac-bd)^2+(bc+ad)^2.$$

Illustration 10. *The determinant of cofactors.* Let

$$\Delta \equiv \begin{vmatrix} a_1 & b_1 & c_1 \\ a_2 & b_2 & c_2 \\ a_3 & b_3 & c_3 \end{vmatrix}$$

be a given determinant and Δ' the determinant in which $a_1, b_1, \ldots$ are replaced by their cofactors $A_1, B_1, \ldots$. For convenience, Δ' is written with rows and columns interchanged in the form

$$\Delta' \equiv \begin{vmatrix} A_1 & A_2 & A_3 \\ B_1 & B_2 & B_3 \\ C_1 & C_2 & C_3 \end{vmatrix}.$$

Then

$$\Delta\Delta' = \begin{vmatrix} a_1A_1+b_1B_1+c_1C_1 & a_1A_2+b_1B_2+c_1C_2 & a_1A_3+b_1B_3+c_1C_3 \\ a_2A_1+b_2B_1+c_2C_1 & a_2A_2+b_2B_2+c_2C_2 & a_2A_3+b_2B_3+c_2C_3 \\ a_3A_1+b_3B_1+c_3C_1 & a_3A_2+b_3B_2+c_3C_2 & a_3A_3+b_3B_3+c_3C_3 \end{vmatrix}.$$

Hence, by the relations on p. 254,

$$\Delta\Delta' = \begin{vmatrix} \Delta & 0 & 0 \\ 0 & \Delta & 0 \\ 0 & 0 & \Delta \end{vmatrix} = \Delta^3,$$

so that
$$\Delta' = \Delta^2.$$

Thus *the determinant of the cofactors is equal (for a determinant of three rows and columns) to the square of the original determinant.*

Illustration 11. *To prove the formula*

$$(a^3+b^3+c^3-3abc)(p^3+q^3+r^3-3pqr) = u^3+v^3+w^3-3uvw,$$

where
$$u = ap+bq+cr, \quad v = ar+bp+cq, \quad w = aq+br+cp.$$

We have seen (p. 268) that

$$a^3+b^3+c^3-3abc = \begin{vmatrix} a & b & c \\ c & a & b \\ b & c & a \end{vmatrix}.$$

Similarly

$$p^3+q^3+r^3-3pqr = \begin{vmatrix} p & q & r \\ r & p & q \\ q & r & p \end{vmatrix} = \begin{vmatrix} p & r & q \\ q & p & r \\ r & q & p \end{vmatrix}.$$

The left-hand side is thus the product

$$\begin{vmatrix} a & b & c \\ c & a & b \\ b & c & a \end{vmatrix} \begin{vmatrix} p & r & q \\ q & p & r \\ r & q & p \end{vmatrix}$$

$$= \begin{vmatrix} ap+bq+cr & ar+bp+cq & aq+br+cp \\ cp+aq+br & cr+ap+bq & cq+ar+bp \\ bp+cq+ar & br+cp+aq & bq+cr+ap \end{vmatrix}$$

$$= \begin{vmatrix} u & v & w \\ w & u & v \\ v & w & u \end{vmatrix}$$

$$= u^3+v^3+w^3-3uvw.$$

***Illustration* 12.** *To prove the identity*

$$\begin{vmatrix} (x-a)^2 & (x-b)^2 & (x-c)^2 \\ (y-a)^2 & (y-b)^2 & (y-c)^2 \\ (z-a)^2 & (z-b)^2 & (z-c)^2 \end{vmatrix} = 2(b-c)(c-a)(a-b)(y-z)(z-x)(x-y).$$

By the method given on p. 265, we have the relations

$$\begin{vmatrix} x^2 & -2x & 1 \\ y^2 & -2y & 1 \\ z^2 & -2z & 1 \end{vmatrix} = 2(y-z)(z-x)(x-y),$$

$$\begin{vmatrix} 1 & 1 & 1 \\ a & b & c \\ a^2 & b^2 & c^2 \end{vmatrix} = (b-c)(c-a)(a-b).$$

Now $\begin{vmatrix} x^2 & -2x & 1 \\ y^2 & -2y & 1 \\ z^2 & -2z & 1 \end{vmatrix} \times \begin{vmatrix} 1 & 1 & 1 \\ a & b & c \\ a^2 & b^2 & c^2 \end{vmatrix}$

$$= \begin{vmatrix} x^2-2ax+a^2 & x^2-2bx+b^2 & x^2-2cx+c^2 \\ y^2-2ay+a^2 & y^2-2by+b^2 & y^2-2cy+c^2 \\ z^2-2az+a^2 & z^2-2bz+b^2 & z^2-2cz+c^2 \end{vmatrix},$$

which is the left-hand side of the required relation, and the result is immediate.

Illustration **13.** *To prove that, if the equation*

$$ax^2+2hxy+by^2+2gx+2fy+c = 0$$

represents two straight lines, then

$$\begin{vmatrix} a & h & g \\ h & b & f \\ g & f & c \end{vmatrix} = 0.$$

If the lines are

$$u_1x+v_1y+w_1 = 0, \quad u_2x+v_2y+w_2 = 0,$$

then the left-hand side of the equation may be equated identically to the product

$$(u_1x+v_1y+w_1)(u_2x+v_2y+w_2),$$

one linear factor for each line. Hence

$$u_1u_2 = a, \quad v_1v_2 = b, \quad w_1w_2 = c,$$

$$v_1w_2+v_2w_1 = 2f, \quad w_1u_2+w_2u_1 = 2g, \quad u_1v_2+u_2v_1 = 2h.$$

Now form the product

$$\begin{vmatrix} u_1 & u_2 & 0 \\ v_1 & v_2 & 0 \\ w_1 & w_2 & 0 \end{vmatrix} \begin{vmatrix} u_2 & v_2 & w_2 \\ u_1 & v_1 & w_1 \\ 0 & 0 & 0 \end{vmatrix}$$

$$= \begin{vmatrix} u_1u_2+u_2u_1 & u_1v_2+u_2v_1 & u_1w_2+u_2w_1 \\ v_1u_2+v_2u_1 & v_1v_2+v_2v_1 & v_1w_2+v_2w_1 \\ w_1u_2+w_2u_1 & w_1v_1+w_2v_1 & w_1w_2+w_2w_1 \end{vmatrix}$$

$$= \begin{vmatrix} 2a & 2h & 2g \\ 2h & 2b & 2f \\ 2g & 2f & 2c \end{vmatrix}.$$

But the left-hand side is the product of two determinants, each of which is zero. Hence, dividing the right-hand side by 8,

$$\begin{vmatrix} a & h & g \\ h & b & f \\ g & f & c \end{vmatrix} = 0.$$

Examples 8

1. Prove that, if

$$ax+by+cz = 1,$$
$$cx+ay+bz = 0,$$
$$bx+cy+az = 0,$$

then

$$\begin{vmatrix} x & y & z \\ z & x & y \\ y & z & x \end{vmatrix} \times \begin{vmatrix} a & b & c \\ c & a & b \\ b & c & a \end{vmatrix} = 1.$$

2. By squaring the determinant

$$\begin{vmatrix} 1 & 1 & 1 \\ a & b & c \\ a^2 & b^2 & c^2 \end{vmatrix},$$

prove that

$$\begin{vmatrix} 3 & a+b+c & a^2+b^2+c^2 \\ a+b+c & a^2+b^2+c^2 & a^3+b^3+c^3 \\ a^2+b^2+c^2 & a^3+b^3+c^3 & a^4+b^4+c^4 \end{vmatrix} = (b-c)^2(c-a)^2(a-b)^2.$$

3. Verify that
$$\begin{vmatrix} a & b & c \\ c & a & b \\ b & c & a \end{vmatrix} = -\begin{vmatrix} a & c & b \\ c & b & a \\ b & a & c \end{vmatrix},$$

and deduce that

$$\begin{vmatrix} a & b & c \\ c & a & b \\ b & c & a \end{vmatrix}^2 = \begin{vmatrix} 2bc-a^2 & c^2 & b^2 \\ c^2 & 2ca-b^2 & a^2 \\ b^2 & a^2 & 2ab-c^2 \end{vmatrix}.$$

4. Given that
$$\begin{vmatrix} a & b & c \\ b & c & a \\ c & a & b \end{vmatrix}$$

is a factor of
$$\begin{vmatrix} 2ab & ac+b^2 & bc+a^2 \\ ac+b^2 & 2bc & ab+c^2 \\ bc+a^2 & ab+c^2 & 2ac \end{vmatrix},$$

find the other factor as a determinant of the third order.

5. If
$$s_n \equiv \alpha^n+\beta^n+\gamma^n,$$

express
$$\begin{vmatrix} s_0 & s_2 & s_3 \\ s_4 & s_6 & s_7 \\ s_6 & s_8 & s_9 \end{vmatrix}$$

as a product of two determinants of the third order.

6. By multiplying the determinants

$$\begin{vmatrix} a^2 & a & 1 \\ b^2 & b & 1 \\ c^2 & c & 1 \end{vmatrix}, \quad \begin{vmatrix} 1 & 1 & 1 \\ a & b & c \\ 0 & 0 & 0 \end{vmatrix},$$

prove that
$$\begin{vmatrix} 2a^2 & b^2+ba & c^2+ca \\ a^2+ab & 2b^2 & c^2+cb \\ a^2+ac & b^2+bc & 2c^2 \end{vmatrix} = 0.$$

7. Express $\begin{vmatrix} 2xy & x+y^2 & x^2+y \\ x^2+y & 2xy & x+y^2 \\ x+y^2 & x^2+y & 2xy \end{vmatrix} \div \begin{vmatrix} x & y & 0 \\ y & 0 & x \\ 0 & x & y \end{vmatrix}$

as a determinant of the third order.

8. Prove that, if

$$x = a^2+2bc, \quad y = b^2+2ca, \quad z = c^2+2ab,$$

$$s = a^2+b^2+c^2, \quad p = bc+ca+ab,$$

then $$\begin{vmatrix} a & b & c \\ c & a & b \\ b & c & a \end{vmatrix}^2 = \begin{vmatrix} x & z & y \\ y & x & z \\ z & y & x \end{vmatrix} = \begin{vmatrix} s & p & p \\ p & s & p \\ p & p & s \end{vmatrix}.$$

9. The determinant

$$\begin{vmatrix} x & a & b & c \\ -a & x & c & -b \\ -b & -c & x & a \\ -c & b & -a & x \end{vmatrix}$$

is denoted by $D(x)$. Prove that

$$D(x).D(-x) = (x^2+a^2+b^2+c^2)^4,$$

and deduce that $$D(x) = (x^2+a^2+b^2+c^2)^2.$$

10. Given that

$$D = \begin{vmatrix} 1 & \omega & \omega^2 & 1 \\ \omega & \omega^2 & 1 & 1 \\ \omega^2 & 1 & 1 & \omega \\ 1 & 1 & \omega & \omega^2 \end{vmatrix},$$

where $\omega^3 = 1$, $\omega \neq 1$, prove that

$$D^2 = -27.$$

Revision Examples XV

1. Evaluate the determinants

(i) $\begin{vmatrix} 1 & 2 & 4 \\ 2 & 4 & 8 \\ 4 & 8 & 16 \end{vmatrix}$, (ii) $\begin{vmatrix} 8 & -7 & 6 \\ -5 & 4 & -3 \\ 2 & -1 & 0 \end{vmatrix}$,

(iii) $\begin{vmatrix} 101 & 102 & 103 \\ 85 & 87 & 89 \\ 61 & 64 & 67 \end{vmatrix}$, (iv) $\begin{vmatrix} 11 & -9 & 13 \\ 15 & 0 & 17 \\ 13 & 19 & -5 \end{vmatrix}$.

Multiply them in pairs according to the 'matrix rule' (p. 268), and check that your answers agree.

2. Evaluate the determinants

(i) $\begin{vmatrix} 1 & 2 & 4 & 8 \\ 2 & 6 & 18 & 54 \\ 4 & 18 & 81 & 324 \\ 8 & 54 & 324 & 0 \end{vmatrix}$, (ii) $\begin{vmatrix} 3 & 5 & 7 & 9 \\ -1 & 2 & -3 & 4 \\ 9 & -7 & 5 & -3 \\ 4 & 3 & 2 & 1 \end{vmatrix}$,

(iii) $\begin{vmatrix} 108 & 40 & 112 & 212 \\ -45 & 20 & -47 & -88 \\ 109 & 40 & 113 & 214 \\ 110 & 30 & 108 & 216 \end{vmatrix}$, (iv) $\begin{vmatrix} 3 & 7 & 1 & 6 \\ 36 & -20 & 8 & -12 \\ 3 & -4 & 1 & 1 \\ -6 & 1 & 2 & -8 \end{vmatrix}$,

(v) $\begin{vmatrix} 1 & 2 & 3 & 4 \\ 3 & -5 & 7 & -9 \\ 7 & 10 & 13 & 16 \\ 13 & -17 & 21 & -25 \end{vmatrix}$.

3. Prove that, in a triangle ABC,

$$\begin{vmatrix} \sin 2A & \sin C & \sin B \\ \sin C & \sin 2B & \sin A \\ \sin B & \sin A & \sin 2C \end{vmatrix} = 0.$$

4. Prove that, if a, b, c are unequal, the non-zero solutions of the equation

$$\begin{vmatrix} x^2-a^2 & x^2-b^2 & x^2-c^2 \\ (x-a)^3 & (x-b)^3 & (x-c)^3 \\ (x+a)^3 & (x+b)^3 & (x+c)^3 \end{vmatrix} = 0$$

are

$$\pm\sqrt{\{(bc+ca+ab)/3\}}.$$

5. Show that, if x is added to all the elements of any determinant, the resulting determinant has the value $A+Bx$, where A and B are independent of x.

6. Prove that

$$\begin{vmatrix} 1+a_1 & a_2 & a_3 \\ a_1 & 1+a_2 & a_3 \\ a_1 & a_2 & 1+a_3 \end{vmatrix} = 1+a_1+a_2+a_3.$$

Deduce, or prove otherwise, that

$$\begin{vmatrix} x_1 & a_1 & a_1 \\ a_2 & x_2 & a_2 \\ a_3 & a_3 & x_3 \end{vmatrix} = \left(1+\sum_{r=1}^{3}\frac{a_r}{x_r-a_r}\right)(x_1-a_1)(x_2-a_2)(x_3-a_3).$$

7. By considering the determinant of cofactors for

$$\begin{vmatrix} a & b & c \\ c & a & b \\ b & c & a \end{vmatrix},$$

or otherwise, prove that

$$\begin{vmatrix} a^2-bc & b^2-ca & c^2-ab \\ c^2-ab & a^2-bc & b^2-ca \\ b^2-ca & c^2-ab & a^2-bc \end{vmatrix} = (a^3+b^3+c^3-3abc)^2.$$

8. Show that, if α, β are the roots of the quadratic

$$ax^2+2bx+c = 0,$$

then

$$\begin{vmatrix} x^2 & x & 1 \\ \alpha^2 & \alpha & 1 \\ \beta^2 & \beta & 1 \end{vmatrix} = D(ax^2+2bx+c),$$

where D is independent of x.

9. Prove that

$$\begin{vmatrix} 1+x^2 & x & 0 \\ x & 1+x^2 & x \\ 0 & x & 1+x^2 \end{vmatrix} = 1+x^2+x^4+x^6.$$

10. Prove that the sum of the roots of the equation

$$\begin{vmatrix} x & 1 & 2 \\ 1 & x & 3 \\ 2 & 3 & x \end{vmatrix} = 0$$

is zero, and that the sum of the squares of the roots is 28.

11. Solve the equation

$$\begin{vmatrix} a-x & b & b \\ b & a-x & b \\ b & b & a-x \end{vmatrix} = 0.$$

12. Prove that

$$\begin{vmatrix} \cos(A-\alpha) & \cos(A-\beta) & \cos(A-\gamma) \\ \cos(B-\alpha) & \cos(B-\beta) & \cos(B-\gamma) \\ \cos(C-\alpha) & \cos(C-\beta) & \cos(C-\gamma) \end{vmatrix} = 0.$$

13. Evaluate the determinant

$$\begin{vmatrix} \cos\theta\cos\psi & \cos\theta\sin\psi & \sin\theta \\ \sin\theta\cos\psi & \sin\theta\sin\psi & -\cos\theta \\ \cos\theta\sin\psi & -\cos\theta\cos\psi & 0 \end{vmatrix}.$$

14. Prove that, if $a^2 = bc$, then

$$\begin{vmatrix} 1 & 1 & a \\ a^2 & ab & b^3 \\ a^2 & ac & c^3 \end{vmatrix} = \begin{vmatrix} 1 & a^2 & a^3 \\ 1 & b^2 & b^3 \\ 1 & c^2 & c^3 \end{vmatrix}.$$

15. Prove that

$$\begin{vmatrix} 1 & \cos\alpha & \sin\alpha \\ 1 & \cos\beta & \sin\beta \\ 1 & \cos\gamma & \sin\gamma \end{vmatrix} = 4\sin\frac{\beta-\gamma}{2}\sin\frac{\gamma-\alpha}{2}\sin\frac{\alpha-\beta}{2}.$$

16. Solve the equation

$$\begin{vmatrix} x^3 & x & 1 \\ 8 & 2 & 1 \\ 27 & 3 & 1 \end{vmatrix} = 0.$$

17. Solve the equation

$$\begin{vmatrix} (x-1)^3 & (x+1)^3 & 8 \\ 1 & 8 & 1 \\ 8 & 27 & 1 \end{vmatrix} = 0.$$

18. Prove that

$$\begin{vmatrix} (b+c)^2 & a^2 & a^2 \\ b^2 & (c+a)^2 & b^2 \\ c^2 & c^2 & (a+b)^2 \end{vmatrix} = 2abc(a+b+c)^3.$$

19. Solve the equation

$$\begin{vmatrix} a & b & x \\ a^2 & b^2 & x^2 \\ a^4 & b^4 & x^4 \end{vmatrix} = 0,$$

where a, b are unequal and not zero.

20. Show that
$$\begin{vmatrix} 1 & 1 & 1 \\ 1 & 1+a & 1 \\ 1 & 1 & 1+a \end{vmatrix} = a^2$$

and that
$$\begin{vmatrix} 1 & 1 & 1 & 1 \\ 1 & 1+a & 1 & 1 \\ 1 & 1 & 1+a & 1 \\ 1 & 1 & 1 & 1+a \end{vmatrix} = a^3.$$

21. Evaluate as a product of factors

(i) $\begin{vmatrix} 1 & b+c & (b-c)(b^2-c^2) \\ 1 & c+a & (c-a)(c^2-a^2) \\ 1 & a+b & (a-b)(a^2-b^2) \end{vmatrix}$,

(ii) $\begin{vmatrix} a & b & c \\ b+c & c+a & a+b \\ a^3-abc & b^3-abc & c^3-abc \end{vmatrix}$.

22. Prove that

$$\begin{vmatrix} a^2 & a^2-(b-c)^2 & bc \\ b^2 & b^2-(c-a)^2 & ca \\ c^2 & c^2-(a-b)^2 & ab \end{vmatrix} = (b-c)(c-a)(a-b)(a+b+c)(a^2+b^2+c^2).$$

23. Prove that

$$\begin{vmatrix} b+c & a-b & a \\ c+a & b-c & b \\ a+b & c-a & c \end{vmatrix} = 3abc-a^3-b^3-c^3.$$

24. Prove that

$$\begin{vmatrix} a & b & ax+b \\ b & c & bx+c \\ ax+b & bx+c & 0 \end{vmatrix} = (b^2-ac)(ax^2+2bx+c).$$

25. Expand
$$\begin{vmatrix} x & a & b \\ -a & x & c \\ -b & -c & x \end{vmatrix}$$
as a polynomial in x.

26. Find a, b such that

$$\begin{vmatrix} x & a & b \\ b & x & a \\ a & b & x \end{vmatrix} \equiv x^3-18x+35.$$

27. Solve the equation

$$\begin{vmatrix} x^3 & a^2+1 & 1 \\ a^3 & x^2+1 & 1 \\ 1 & x^2+a^2 & 1 \end{vmatrix} = 0,$$

where $a^2 \neq 1$.

28. Solve the equation

$$\begin{vmatrix} 2 & x-1 & x+3 \\ x-2 & 6 & x+7 \\ x+2 & x+9 & 28 \end{vmatrix} = 0.$$

29. Prove that

$$\begin{vmatrix} \cos A & \cos B & \cos C \\ \cos^2 A & \cos^2 B & \cos^2 C \\ \sin^2 A & \sin^2 B & \sin^2 C \end{vmatrix}$$

$$= (\cos B - \cos C)(\cos C - \cos A)(\cos A - \cos B).$$

30. Prove that, if x, y, z are different and

$$\begin{vmatrix} x & x^3 & x^2+1 \\ y & y^3 & y^2+1 \\ z & z^3 & z^2+1 \end{vmatrix} = 0,$$

then

$$xyz = x+y+z.$$

31. Express as products of factors

(i)
$$\begin{vmatrix} x & y & z \\ x(y+z) & y(z+x) & z(x+y) \\ (y+z)^2 & (z+x)^2 & (x+y)^2 \end{vmatrix},$$

(ii)
$$\begin{vmatrix} x & y & z \\ x^3 & y^3 & z^3 \\ y^3+z^3 & z^3+x^3 & x^3+y^3 \end{vmatrix}.$$

32. Find four linear factors of the determinant

$$\begin{vmatrix} 0 & a & b & c \\ a & 0 & c & b \\ b & c & 0 & a \\ c & b & a & 0 \end{vmatrix}.$$

33. Solve the equation

$$\begin{vmatrix} x^4 & x^2 & x & 1 \\ 16 & 4 & 2 & 1 \\ 81 & 9 & 3 & 1 \\ 625 & 25 & -5 & 1 \end{vmatrix} = 0.$$

34. Solve the equation

$$\begin{vmatrix} x & b & c & d \\ b & x & c & d \\ b & c & x & d \\ b & c & d & x \end{vmatrix} = 0.$$

35. Solve the equation

$$\begin{vmatrix} a-x & b & c & d \\ a & b-x & c & d \\ a & b & c-x & d \\ a & b & c & d-x \end{vmatrix} = 0.$$

Miscellaneous Examples

1. Prove that, if

$$3yz+2y+z+1 = 0 \quad \text{and} \quad 3zx+2z+x+1 = 0,$$

then
$$3xy+2x+y+1 = 0.$$

2. The three numbers a, b, c, none of which are zero, are related by the equations

$$a^2 = b^2+c^2-2bc\sqrt{(1-a^2)},$$
$$b^2 = c^2+a^2-2ca\sqrt{(1-b^2)},$$
$$c^2 = a^2+b^2-2ab\sqrt{(1-c^2)}.$$

Deduce that
$$a = c\sqrt{(1-b^2)}+b\sqrt{(1-c^2)},$$
$$b = a\sqrt{(1-c^2)}+c\sqrt{(1-a^2)},$$
$$c = b\sqrt{(1-a^2)}+a\sqrt{(1-b^2)},$$

and show also that

$$4a^2b^2c^2 = (a+b+c)(-a+b+c)(a-b+c)(a+b-c).$$

3. If a, b, c are the roots of the equation

$$x^3-px^2+qx-r = 0,$$

show that
$$a^3+b^3+c^3-3abc = p^3-3pq.$$

By forming the equation whose roots are the squares of the roots of the given equation, show that

$$a^6+b^6+c^6-3a^2b^2c^2 = (p^2-2q)(p^4-4p^2q+6pr+q^2).$$

If
$$a^6+b^6+c^6-3a^2b^2c^2 = 0$$

but
$$a^3+b^3+c^3-3abc \neq 0,$$

evaluate a, b in terms of c, where a, b, c are assumed to be real.

4. Prove, by induction or otherwise, that

$$9^{2n}-80n-1$$

is a multiple of 6400 when n is a positive integer greater than 1.

5. Show that the substitution $x = y+h$ in the equation

$$x^4+px^3+qx^2+rx+s = 0$$

will, for suitably chosen h, give an equation involving only y^4, y^2, and a constant term, provided that

$$p^3-4pq+8r = 0.$$

Solve the equation $\quad x^4-4x^3+x^2+6x+2 = 0.$

6. The equation

$$f(x) \equiv x^4-10x^3+(33-a^2)x^2-(44-6a^2)x+20-5a^2 = 0$$

has four real roots; two of them are independent of a. Find all the roots and draw sketches of the function $f(x)$ for the values 0, $\frac{1}{2}$, 1, 2, 3 of a, marking clearly the points corresponding to double roots of the equation $f(x) = 0$.

7. Show that there is a value of k for which

$$x^6-15x^3-8x^2+2$$

is divisible by x^2+kx+1, and determine this value.

8. Prove that the function

$$y = \frac{(x-\alpha)^2}{x-\beta}$$

can take (twice) all values except those in an interval of length four times the numerical values of $\alpha-\beta$, and locate this range of values precisely.

9. Prove that, if a, b are positive and $a+b = 1$, then

$$\left(a+\frac{1}{a}\right)^2+\left(b+\frac{1}{b}\right)^2 \geqslant 12\tfrac{1}{2}.$$

10. Find $\quad \displaystyle\sum_{r=1}^{n} \frac{2r+1}{r^2(r+1)^2}.$

11. Given that the equation

$$x^4+6bx^2+4cx+d = 0 \qquad (b,\ c,\ d \text{ real}),$$

has one real double root, prove that for the other two roots to be real and distinct it is necessary that $9b^2 > d$.

12. Find (in terms of a, b, c) the values of x, y, z which satisfy the simultaneous equations

$$(1+a)x+y+z = 1,$$
$$x+(1+b)y+z = 1,$$
$$x+y+(1+c)z = 1.$$

Show also that no (finite) values of x, y, z can be found to satisfy the equations if a, b, c are the roots of a cubic equation of the form

$$At^3+Bt^2+t+1 = 0.$$

13. Prove that, if
$$\alpha+\beta+\gamma = 0,$$
$$\alpha^2+\beta^2+\gamma^2 = Q,$$
$$\alpha^3+\beta^3+\gamma^3 = -R,$$
then α, β, γ are the roots of the equation

$$x^3-\tfrac{1}{2}Qx+\tfrac{1}{3}R = 0.$$

Find, in terms of Q, R, the values of $\alpha^4+\beta^4+\gamma^4$ and $\alpha^5+\beta^5+\gamma^5$.

14. By using partial fractions and the logarithmic series for $\log_e(1-x)$, obtain the sums of the series

$$\sum_{n=1}^{\infty} \frac{x^n}{n(n+1)}, \quad \sum_{n=1}^{\infty} \frac{x^n}{n(n+1)(n+2)},$$

when $-1 \leqslant x < 1$.

Find the sum of the first N terms of the second series when $x = 1$, and deduce that

$$\sum_{n=1}^{\infty} \frac{1}{n(n+1)(n+2)} = \frac{1}{4}.$$

15. Solve the equation

$$8x^6-54x^5+109x^4-108x^3+109x^2-54x+8 = 0.$$

16. Show that the equation

$$x^3+3bx+c = 0$$

can be expressed in the form

$$\beta(x-\alpha)^3 = \alpha(x-\beta)^3,$$

where α, β (assumed distinct) are the roots of the quadratic

$$by^2+cy-b^2=0.$$

17. Prove that, if the polynomial

$$ax^3+x^2-3bx+3b^2$$

has two coincident zeros, then it is, in general, a perfect cube.

Hence, or otherwise, show that if

$$x^4+4ax^3+2x^2-4bx+3b^2$$

has three equal zeros, then the fourth is identical with them, and find for what values of a, b this is the case.

18. Three roots of the quartic equation

$$(x^2+1)^2 = ax(1-x^2)+b(1-x^4)$$

satisfy the equation $\quad x^3+px^2+qx+r = 0.$

Prove that $\quad p^2-q^2-r^2+1 = 0.$

19. The roots $\alpha_1, \alpha_2, \ldots, \alpha_n$ of the polynomial equation $f(x) = 0$ are real and distinct, and

$$\alpha_1 < \alpha_2 < \ldots < \alpha_n.$$

If $f'(\alpha_1)$ is positive, find the sign of $f'(\alpha_r)$.

Prove that the roots of the equation

$$f(x)+\lambda f'(x) = 0,$$

where λ is a real constant, are real and distinct.

20. Find the sum to infinity of the series whose nth term is

$$\frac{1+x+x^2+\ldots+x^{n-1}}{1+2+3+\ldots+n},$$

where $|x| < 1$.

21. If the roots of the equation

$$x^3+px^2+qx+r = 0$$

are in arithmetical progression, find a relation between p, q, r.

State and prove the converse theorem.

If p, q, r are all positive, deduce that $2q^3$ is not less than $27r^2$.

22. Prove that, if α, β, γ are the roots of the equation

$$x^3-3px^2-3(1-p)x+1 = 0,$$

then $\quad \beta(1-\gamma) = \gamma(1-\alpha) = \alpha(1-\beta) = 1$

or $\quad \beta(1-\alpha) = \gamma(1-\beta) = \alpha(1-\gamma) = 1.$

Show that, if p is real, then α, β, γ are real.

23. A square $ABCD$ is divided into 25 squares by two sets each of four equidistant lines. Show that the number of paths of length $2AB$ connecting A to C, and made up of segments of the dividing lines or the edges of the square, is 252.

24. If a, b, c, d are all real, and if

$$(a^2+b^2+c^2)(b^2+c^2+d^2) = (ab+bc+cd)^2,$$

prove that a, b, c, d are in geometrical progression.

25. Prove that, if $f(x)$ is a quadratic polynomial, then

$$\frac{f(x)}{(x-a)(x-b)(x-c)} = \begin{vmatrix} 1 & 1 & 1 \\ a & b & c \\ \dfrac{f(a)}{x-a} & \dfrac{f(b)}{x-b} & \dfrac{f(c)}{x-c} \end{vmatrix} \div \begin{vmatrix} 1 & 1 & 1 \\ a & b & c \\ a^2 & b^2 & c^2 \end{vmatrix}.$$

26. The roots of the equation

$$x^4 = 3px+q$$

are $\alpha, \beta, \gamma, \delta$. Express

$$\begin{vmatrix} 1 & \alpha^4 & \alpha^5 & \alpha^6 \\ 1 & \beta^4 & \beta^5 & \beta^6 \\ 1 & \gamma^4 & \gamma^5 & \gamma^6 \\ 1 & \delta^4 & \delta^5 & \delta^6 \end{vmatrix} \div \begin{vmatrix} 1 & \alpha & \alpha^2 & \alpha^3 \\ 1 & \beta & \beta^2 & \beta^3 \\ 1 & \gamma & \gamma^2 & \gamma^3 \\ 1 & \delta & \delta^2 & \delta^3 \end{vmatrix}$$

in terms of p and q.

27. Express

$$\begin{vmatrix} 1 & 1 & x & x \\ x & x & 1 & x \\ y & x & xy & xy \\ 1 & x & y & 1 \end{vmatrix}$$

as the product of three factors.

28. Prove that, if $\lambda^3 = 1$, one root of the equation

$$\begin{vmatrix} a-x & b & c \\ c & a-x & b \\ b & c & a-x \end{vmatrix} = 0$$

is $x = a+\lambda b+\lambda^2 c$, and hence solve the equation.

29. Prove that, if $\alpha, \beta, \gamma, \delta$ are the roots of the equation

$$px^4+qx^3+rx^2+x+1 = 0,$$

then

$$\begin{vmatrix} 1+\alpha & 1 & 1 & 1 \\ 1 & 1+\beta & 1 & 1 \\ 1 & 1 & 1+\gamma & 1 \\ 1 & 1 & 1 & 1+\delta \end{vmatrix} = 0.$$

30. Express

$$\begin{vmatrix} (b+c)^2 & a^2 & a^2 \\ b^2 & (c+a)^2 & b^2 \\ c^2 & c^2 & (a+b)^2 \end{vmatrix}$$

in terms of $p \equiv a+b+c$, $q \equiv bc+ca+ab$, $r \equiv abc$.

31. Prove that, if a, b, c are the sides of a triangle of area Δ, then

$$\begin{vmatrix} (b-c)^2 & b^2 & c^2 & 1 \\ a^2 & (c-a)^2 & c^2 & 1 \\ a^2 & b^2 & (a-b)^2 & 1 \\ 1 & 1 & 1 & 0 \end{vmatrix} = -16\Delta^2.$$

32. Find all the values of x in the range $0 \leqslant x \leqslant \frac{1}{2}\pi$ for which

$$\begin{vmatrix} \cos^2 2x & \cos^2 4x & \cos^2 6x \\ \sin 2x & \sin 4x & \sin 6x \\ 1 & 1 & 1 \end{vmatrix} = 0.$$

33. Prove that the value of

$$\begin{vmatrix} x & x+a+b+c & x+a+b+c & x+a+b+c \\ y & y-a & y-b & y-c \\ z & z-b & z-c & z-a \\ t & t-c & t-a & t-b \end{vmatrix}$$

is $(x+y+z+t)(a^3+b^3+c^3-3abc)$.

34. Without expanding the determinant, prove that $x+a+b$ is a factor of

$$\begin{vmatrix} x & 0 & a & b \\ b & x & 0 & a \\ a & b & x & 0 \\ 0 & a & b & x \end{vmatrix}.$$

Expand the determinant, and factorize the expression obtained.

35. Evaluate the determinant

$$\begin{vmatrix} 0 & a & b & c \\ -a & 0 & d & e \\ -b & -d & 0 & f \\ -c & -e & -f & 0 \end{vmatrix},$$

expressing your answer as the square of a polynomial in a, b, c, d, e, f.

36. Solve the equation

$$\begin{vmatrix} 0 & x & x^2 & x^3 \\ -x & 0 & -1 & -2 \\ -x^2 & 1 & 0 & -1 \\ -x^3 & 2 & 1 & 0 \end{vmatrix} = 0.$$

37. Prove that, if $x+y+z = 0$, then

$$\begin{vmatrix} 0 & x & -y & -z \\ -x & 0 & xy-z^2 & y^2-zx \\ y & z^2-xy & 0 & x^2-yz \\ z & zx-y^2 & yz-x^2 & 0 \end{vmatrix} = 0.$$

38. Prove that, if $\qquad s_r \equiv \alpha^r+\beta^r+\gamma^r,$

then

$$\begin{vmatrix} s_0 & s_1 & s_2 \\ s_1 & s_2 & s_3 \\ s_2 & s_3 & s_4 \end{vmatrix} \times \begin{vmatrix} s_0 & s_1 & s_3 \\ s_2 & s_3 & s_5 \\ s_4 & s_5 & s_7 \end{vmatrix} = \begin{vmatrix} s_0 & s_1 & s_2 \\ s_1 & s_2 & s_3 \\ s_3 & s_4 & s_5 \end{vmatrix} \times \begin{vmatrix} s_0 & s_1 & s_2 \\ s_2 & s_3 & s_4 \\ s_4 & s_5 & s_6 \end{vmatrix}.$$

39. Show that

$$\begin{vmatrix} a^2+1 & ab & ac & ad \\ ba & b^2+1 & bc & bd \\ ca & cb & c^2+1 & cd \\ da & db & dc & d^2+1 \end{vmatrix} = \begin{vmatrix} a^2+1 & b^2 & c^2 & d^2 \\ a^2 & b^2+1 & c^2 & d^2 \\ a^2 & b^2 & c^2+1 & d^2 \\ a^2 & b^2 & c^2 & d^2+1 \end{vmatrix}$$

$$= 1+a^2+b^2+c^2+d^2.$$

40. Show that

$$\begin{vmatrix} 4 & x+1 & x+1 & x+1 \\ x+1 & x^2-x+1 & x & x \\ x+1 & x & x^2-x+1 & x \\ x+1 & x & x & x^2-x+1 \end{vmatrix} = (x-1)^6.$$

41. Prove that, if $s = a_1+a_2+a_3+a_4$ and $\lambda_i = s-a_i$, then

$$\begin{vmatrix} x-\lambda_1 & a_2 & a_3 & a_4 \\ a_1 & x-\lambda_2 & a_3 & a_4 \\ a_1 & a_2 & x-\lambda_3 & a_4 \\ a_1 & a_2 & a_3 & x-\lambda_4 \end{vmatrix} = x(x-s)^3.$$

42. Prove that the coefficient of x^n in the expansion of

$$\frac{x^2}{(1-ax)(1-bx)(1-cx)}$$

may be written in the form

$$\begin{vmatrix} a^n & b^n & c^n \\ a & b & c \\ 1 & 1 & 1 \end{vmatrix} \div \begin{vmatrix} a^2 & b^2 & c^2 \\ a & b & c \\ 1 & 1 & 1 \end{vmatrix}.$$

43. Prove that

$$\begin{vmatrix} 1 & 1 & 1 & 1 \\ 1 & r & r^2 & r^3 \\ 1 & r^2 & r^4 & r^6 \\ 1 & r^3 & r^6 & r^9 \end{vmatrix} = r^4(r-1)^3(r^2-1)^2(r^3-1).$$

44. Prove that

$$\begin{vmatrix} 1+t_1^2 & 2t_1 & 1-t_1^2 \\ 1+t_2^2 & 2t_2 & 1-t_2^2 \\ 1+t_3^2 & 2t_3 & 1-t_3^2 \end{vmatrix} = -4(t_2-t_3)(t_3-t_1)(t_1-t_2),$$

and deduce that

$$\begin{vmatrix} 1 & \sin\alpha & \cos\alpha \\ 1 & \sin\beta & \cos\beta \\ 1 & \sin\gamma & \cos\gamma \end{vmatrix} = -4\sin\tfrac{1}{2}(\beta-\gamma)\sin\tfrac{1}{2}(\gamma-\alpha)\sin\tfrac{1}{2}(\alpha-\beta).$$

ANSWERS TO EXAMPLES

CHAPTER 1

EXAMPLES 1

1. $\frac{1}{5}$. **2.** $\frac{1}{7}$. **3.** $\frac{23}{13}$ **4.** 1, 2. **5.** 1, 3.

6. $\frac{4}{7}$ **7.** $-6a$. **8.** $a+b$. **9.** 2, 3. **10.** 1, 6.

11. $\frac{7}{5}$. **12.** $-\frac{5}{8}$ **13.** $\frac{9}{16}$. **14.** $-2, 3$.

CHAPTER 2

EXAMPLES 1

1. $3-x$. **2.** $-\frac{3}{2}x$. **3.** $\frac{21}{10}-\frac{2}{5}x$.

4. $1-x$. **5** 3. **6.** $\frac{1}{8}x-\frac{1}{24}$.

7. $a+b-x$. **8.** $x+a$. **9.** 0.

10. $(a+b)\,x-ab$. **11.** x. **12.** $\dfrac{x}{b-a}+\dfrac{b-2a}{b-a}$.

13. $p \neq 0$, $x = 1/p$; $p = 0$, no solution.

14. $p \neq 1$ or 2, $x = 1/\{(p-1)(p-2)\}$; $p = 1$ or 2, no solution

15. $p \neq 0$, $x = 1+(1/p)$; $p = 0$, no solution.

16. $p \neq 2$, $x = q/(p-2)$; $p = 2$, $q \neq 0$, no solution; $p = 2$, $q = 0$, x can have any value.

17. $p \neq 0$, $x = (q+3)/p^2$; $p = 0$, $q \neq -3$, no solution; $p = 0$, $q = -3$, x can have any value.

18. $p \neq 0$ or -4, $x = (q+5)/\{p(p+4)\}$; $p = 0$ or -4, $q \neq -5$, no solution; $p = 0$ or -4, $q = -5$, x can have any value.

19. $a \neq b$ or $-b$, $x = 1/(a^2-b^2)$; $a = b$ or $-b$, no solution.

20. $a \neq 2$ or 4; $x = (b+6)/\{(a-2)(a-4)\}$; $a = 2$ or 4, $b \neq -6$, no solution; $a = 2$ or 4, $b = -6$, x can have any value.

CHAPTER 3

EXAMPLES 1

1. x^2-3x+2. **2.** x^2+x+1. **3.** $2x^2+3x+4$.

4. x^2-5x+4. **5.** $\frac{1}{6}x^2+\frac{5}{6}x+1$. **6.** $3x^2-2x-1$.

EXAMPLES 2

1. 3·24, −1·24.
2. 2·35, −0·85.
3. 0·10, −9·10.
4. 7·74, 0·26.
5. 2·55, −0·30.
6. 2·78, 0·72.
7. 2·48, −1·48.
8. 2·22, 0·18.

EXAMPLES 3

1. 4.
2. 1.
3. $\pm\frac{9}{2}$.
4. ± 12.
5. ± 24.
6. ± 12.
7. 1.
8. 121.
9. $\frac{2}{9}, -\frac{1}{3}$.

EXAMPLES 4

1. $x^2+7x-18 = 0$.
2. $3x^2-10x+6 = 0$.
3. $25x^2-71x+49 = 0$.
4. $x^2+20x+76 = 0$.
5. Read example 6 with $a = 1, b = m, c = n$.
6. (i) $ax^2+(b-4a)\,x+(c-2b+4a) = 0$.
   (ii) $a^2x^2+a(b-2c)\,x+c(c-b+a) = 0$.
   (iii) $a^2x^2+(2ac-b^2)\,x+c^2 = 0$.
   (iv) $cx^2+bx+a = 0$.
   (v) $acx^2+b(a+c)\,x+(a^2+b^2+c^2-2ac) = 0$.
   (vi) $a^2x^2+(2ac+2a^2-b^2)\,x+(c+a)^2-b^2 = 0$.
   (vii) $a^3x^2+b(b^2-3ac)\,x+c^3 = 0$.
   (viii) $c^2x^2+(2ac-b^2)\,x+a^2 = 0$.
   (ix) $a^2x^2+3abx+2b^2+ac = 0$.
   (x) $a^4x^2+3a^2(2ac-b^2)\,x+2b^4+9a^2c^2-8ab^2c = 0$.

CHAPTER 4

EXAMPLES 1

1. $1, 3; 2\pm\sqrt{3}; 2\pm i$.
2. $-5, -3; -4\pm\sqrt{5}; -4\pm 2i$.
3. $-1, 3; 1\pm\sqrt{5}; 1\pm 3i$.

EXAMPLES 2

1. $10+3i$.
2. $-8+8i$.
3. $34+22i$.
4. $-16-3i$.

5. $2+7i$. **6.** $0+0i$.

7. $0+2i$. **8.** $2+11i$.

9. $-2-16i$. **10.** $-10+0i$.

EXAMPLES 3

1. $-7+22i$. **2.** $26+2i$.

3. $7-i$. **4.** a^2+b^2.

5. $-3+4i$. **6.** $\cos(A+B)+i\sin(A+B)$.

7. 10. **8.** $-46+9i$.

EXAMPLES 4

1. i. **2.** $\frac{24}{25}+\frac{7}{25}i$. **3.** $\frac{5}{7}-\frac{6}{7}i$.

4. $\frac{5}{17}-\frac{14}{17}i$. **5.** $-\frac{27}{37}+\frac{23}{37}i$. **6.** $\cos\theta+i\sin\theta$.

EXAMPLES 5

1. $5-2i$, $-1+8i$, $21-i$, $-\frac{9}{34}+\frac{19}{34}i$.

2. $-7+9i$, $-1-5i$, $-2-34i$, $\frac{13}{29}+\frac{11}{29}i$.

3. $4+2i$, $4-2i$, $8i$, $-2i$. **4.** $3+i$, $3+3i$, $2-3i$, $-2+3i$.

5. 4, $6i$, 13, $-\frac{5}{13}+\frac{12}{13}i$. **6.** -6, $8i$, 25, $-\frac{7}{25}-\frac{24}{25}i$.

7. $\frac{3}{25}-\frac{4}{25}i$. **8.** $-\frac{5}{169}-\frac{12}{169}i$. **9.** $-\frac{1}{6}i$.

10. $-\frac{3}{50}+\frac{2}{25}i$. **11.** $-2\pm 3i$. **12.** $1\pm i$.

13. $-3\pm i$. **14.** $\pm\frac{3}{2}i$. **15.** $4\pm 3i$.

16. $-2\pm i$.

EXAMPLES 7

3. 2, $-30°$; 5, $53°\ 7'$; 13, $112°\ 36'$; 3, $0°$; 10, $-53°\ 7'$; 2, $-90°$.

5. Straight line $4x+10y-21 = 0$.

EXAMPLES 8

1. (*a*) (3, 2); (*b*) (2, 1); (*c*) (4, 7).

2. (i) (*a*) (1, 0); (*b*) (3, -5); (*c*) (3, -5).
(ii) (*a*) (2, -1); (*b*) (2, 1); (*c*) (5, 0).

Chapter 5

EXAMPLES 1

See p. 290.

EXAMPLES 2

1. $x^2+(p-2)x+q-p+1 = 0.$ 3. $x^2+3x+43 = 0.$
5. $9x^2+10x+3 = 0.$ 6. $8x^2+10x+5 = 0.$
7. $x^2-34x+37 = 0.$ 8. $x^2+3x+4 = 0$; 1.
9. $x^2-5x+100 = 0.$ 11. $5x^2+69x+581 = 0.$
12. $x^2-18x+54 = 0.$
14. (i) $a^2x^2+2(ab-2h^2)x+b^2 = 0$; $a^3x^2-4ah^2x+4bh^2 = 0.$
16. α/β, β/α. 18. $\alpha+1/\beta$, $\beta+1/\alpha$. 19. $2\alpha-\beta$, $2\beta-\alpha$.

EXAMPLES 3

1. x^2-x+2; 11. 2. x^2+6x+9; 11.
3. $x^3+7x^2+18x+56$; 169. 4. $x+3$; -10.
5. x^2-4x+3; 0. 6. $3x^2-x+1$; 0.
7. $2x^3-4x^2+9x-18$; 37. 8. $5x^2-15x+46$; -138.
9. $4x^2+6x+12$; 23. 10. $7x^3-14x^2+28x-56$; 95.

EXAMPLES 4

1. 25. 2. 84. 3. 27. 4. -6.
5. 785. 6. 89. 7. 259. 8. -11.

EXAMPLES 5

1. $x^3+9x^2+29x+34 = 0.$ 2. $x^3-6x^2+11x-5 = 0.$
3. $x^4+12x^3+48x^2+73x+31 = 0.$ 4. $2x^3-9x^2+12x-11 = 0.$
5. $3x^2-13x+19 = 0.$ 6. $3x^3+14x^2+19x = 0.$

EXAMPLES 6

1. $x^2+20x+64 = 0.$ 2. $x^3+18x-189 = 0.$
3. $x^4-6x^3-8x-16 = 0.$ 4. $3x^3+2x^2+x+1 = 0.$
5. $2x^4+300x^2-1000x-20{,}000 = 0.$
6. $x^3-50x^2+300x+7000 = 0.$

REVISION EXAMPLES I

1. $-3, 4$. **2.** $1, 0, 1, 1$.

5. $7, -1$. **6.** $-1, -2$. **7.** $7, 17, 17$.

8. $-4a-14$; $-\frac{7}{2}$. **9.** -11; $-1, -1, -\frac{1}{2}, 3$.

10. $-10, 31, -30$; 8. **11.** $15, 75$. **12.** $1, 25$.

13. $2, -8$. **14.** $x+2$. **15.** $-3, 2$.

16. $135, -329$. **17.** $x^4-6x^3+15x^2-20x+12$. **18.** $-25, 22$.

19. $3x^5+4x^4-3x^3-11x^2+4x+12$.

20. $\frac{1}{128}(35x^9-180x^7+378x^5-420x^3+315x)$.

CHAPTER 6

EXAMPLES 1

1. Symmetry. **2.** Nothing. **3.** Symmetry.

4. Skew-symmetry. **5.** Skew-symmetry. **6.** Symmetry.

REVISION EXAMPLES II

1. $6, -2, \pm 2i$. **2.** $3, 6, 12$.

3. $x^3-(3r-3pq+p^3)\,x^2+(3r^2-3pqr+q^3)\,x-r^3 = 0$.

4. $6, 7, -2, -3$. **5.** $3, 4, 6, -2$.

6. $x^3-11x^2+13x-4 = 0$. **7.** $\pm\sqrt{3}, 1\pm i$.

8. $k = -12$; $\pm\sqrt{3}, -2\pm\sqrt{2}$.

9. (i) $x^4-3x^3+5x^2-3x+7 = 0$.
(ii) $x^4+3x^3+14x^2+24x+49 = 0$.

10. $\frac{1}{2}(1\pm\sqrt{13}), \frac{1}{2}(1\pm i\sqrt{7})$.

11. $\pm\frac{5}{3}, \frac{1}{5}(3\pm i\sqrt{6})$. **12.** $\frac{1}{2}, \frac{1}{2}, \frac{1}{2}, -\frac{3}{2}$.

13. $\frac{1}{2}(-1\pm i\sqrt{3}), \frac{1}{2}(-1\pm\sqrt{13})$. **14.** $\frac{1}{2}(3\pm\sqrt{13}), \frac{1}{2}(-5\pm\sqrt{17})$.

15. $-1/k, -(k+1)/(k-1)$; $\frac{1}{2}(1\pm\sqrt{5}), -2\pm\sqrt{5}$.

17. $q^2x^3-2pqx^2+p^2x+q$.

18. $\lambda = -3$; $2\pm\sqrt{2}, 2\pm\sqrt{3}$.

19. $\{(p-1)\,x+q\}^3 = 0$. **20.** -1.

21. $2x^2-3x-2$.
22. 0; 0; 1.

23. $-pq$, p^2q, $-p^3q+\frac{1}{3}q^3$, p^4q-pq^3, where $s_2=-p$, $s_3=q$.

24. $-\frac{7}{5}(\beta\gamma+\gamma\alpha+\alpha\beta)$.
25. 2.

26. $\frac{1}{3}c+\frac{1}{6}a^3-\frac{1}{2}ab$, $\frac{4}{3}ac+\frac{1}{2}b^2+\frac{1}{6}a^4-a^2b$.

28. 1, -1, 2 or 1, -1, -2.

Chapter 7

EXAMPLES 1

1. 1, -2.
2. 4.
3. 2.
4. 0, 2, $-\frac{6}{5}$.

5. 1.
6. 9, $8\frac{1}{5}$.
7. 30.
8. 6.

9. 25. (The solution $x=0$ satisfies, but both sides are then imaginary.)

10. 5.
11. 8.
12. 0, 2, -2.

13. $\frac{1}{2}$.
14. 5, 6.
15. ± 12.

16. $\frac{4}{13}$, $\frac{9}{13}$.
17. 3, $\frac{1}{3}$, -6, $-\frac{1}{6}$.

18. -1, -1, $\frac{1}{2}(3\pm\sqrt{5})$.
19. 1, 1, $-3\pm\sqrt{8}$.

20. $\frac{1}{2}(1\pm i\sqrt{3})$, $\frac{1}{2}(1\pm i\sqrt{3})$, 2, $\frac{1}{2}$.
21. -1, -1, $\frac{1}{2}(-3\pm\sqrt{5})$.

22. 2, $\frac{1}{2}$, -3, $-\frac{1}{3}$.
23. $\frac{1}{2}(-1\pm i\sqrt{3})$, $\frac{1}{2}(-1\pm i\sqrt{3})$.

24. 0, 0, -1, -1. [The complex roots arise from $x^2+x+1=0$.]

25. 1, $\frac{7}{5}$, $-\frac{9}{7}$.
26. 2, 4, -6.
27. 4, $\frac{1}{6}(5\pm\sqrt{13})$.

28. -1, -2, 6.
29. -3, $\frac{1}{2}(3\pm i\sqrt{39})$.
30. 2, $\frac{3}{2}$, $-\frac{4}{3}$.

31. 1, $\frac{1}{2}$, $\frac{1}{3}$.
32. 1, 1, $-\frac{3}{2}$.
33. 1, $\frac{1}{2}$, $-\frac{1}{6}$.

34. 1, -1, 4.
35. -2, -3, $-\frac{1}{4}$.
36. 2, 5, $-\frac{2}{3}$.

37. 1, -1, -2, 4.
38. -1, -4, 2, 3.
39. -2, -5, 3, 5.

40. 2, 2, $-\frac{1}{2}$, $-\frac{1}{2}$.
41. -2, -3, 3, 4.
42. 3, -4, $-\frac{3}{2}$, $-\frac{4}{3}$.

43. 2, $-2\pm\sqrt{3}$, $-3\pm\sqrt{8}$.

44. 7, -4, -4 (and $x^2+x+3=0$ for the others).

Chapter 8

REVISION EXAMPLES III

1. (1, 1), $(\frac{3}{2}, \frac{3}{4})$.
2. $(\frac{3}{8}, -\frac{11}{8})$.

3. $(-1, 1)$, $(-4, 2)$.
4. $(1, -2)$, $(\frac{13}{7}, -\frac{11}{7})$.

5 $(4, 1), (-1, -4)$.

6. $(-1, 2), (-\frac{1}{2}, \frac{7}{4})$.

7. $b = 2, a = 3$; $(1, 2), (\frac{73}{16}, -\frac{131}{72})$.

8. $(1, -1), (-\frac{49}{26}, \frac{12}{13})$.

9. $(5, -2), (-2, 5)$.

10. $(\pm 12, \pm 8, \pm 6)$.

11. $(3, -1, 2)$.

12. $(5, 2), (5, 2)$.

13. $(1, 1), (-1, 2)$.

14. $(4, 3), (3, 4)$.

15. $(2, -1), (-\frac{14}{3}, -21)$.

16. $(5, \frac{6}{5})$.

17. $(4, 3), (-2, -3)$.

18. $(-1, -1), (\frac{19}{23}, \frac{5}{23})$.

19. $(3, -1), (-\frac{21}{4}, \frac{25}{8})$.

20. $(11, 8), (-8, -11)$.

21. $(-11, 6), (-3, 2)$.

22. $(2, -1), (-\frac{1}{4}, \frac{1}{2})$.

23. $(\frac{3}{4}, 4), (\frac{1}{4}, \frac{4}{3})$.

24. $(1, \frac{3}{2}), (3, \frac{1}{2})$.

25. $(\pm 3, \mp 3), (\pm 9\sqrt{\frac{3}{17}}, \pm 6\sqrt{\frac{3}{17}})$.

26. $(\pm 5, \pm 2), (\pm \frac{7}{2}, \pm \frac{1}{2})$.

27. $(\pm 3, \pm 4), (\pm 2, \pm 3)$.

28. $(\pm 4, \pm \frac{3}{2}), (\pm 3, \pm 2)$.

29. $(\pm 1, \pm 2), (\pm \frac{1}{3}, \mp \frac{4}{3})$.

30. $x^2y^2 - 50xy + 264 = 0$; $(2, 3), (3, 2)$.

31 $(-1, 2, -1)$.

32. $(1, 2, 3)$.

33. $(1, 2, 3)$.

34. $(5, 4, 1)$.

35. $(0, 0, 1)$.

36. $(1, -2, 3)$.

37. $(0, 1, 2)$.

38. $(1, -2, 1)$.

Chapter 9

EXAMPLES 1

2. (i) 1·002, 1·0002; (ii) 1·003, 1·0003;
(iii) 0·999, 0·9999; (iv) 0·996, 0·9996.

EXAMPLES 2

1. 1·2, −0·4, −1·8.

2. 0·7.

3. −1·32.

4. 0·4, 3·2, −0·7.

5. 0·7, 2·5, −1·2.

6. 1·9, −1·5, −0·4.

7. 2·1, 0·4, −0·5.

8. 3·9, 1·3, −0·2.

9. 0 9.

10. 0, 2·3, 3 7.

11. 0·4, 3·2, −0·7.

12. 2·7.

13. 1·9, −2·1.

14. 1·8, −0·2, −1·6.

15. 1, 2·7; 0·7, 3·1.

16. 1·1.

17. 0·8.

18. 1·24, (−3·24), 0.

19. 3·27, 2·24.

20. 2·45.

21. 22; 1·8.

REVISION EXAMPLES IV

2. $(x-1)(3x-5)$; $1, \frac{5}{3}$.

3. $4(x-1)(x-2)(x-3)$; $1, 2, 3$; $8 \leqslant n \leqslant 9$.

5. 2. **7.** $8 \leqslant k \leqslant 11$. **8.** $0 \leqslant \lambda \leqslant \frac{4}{27}$.

10. $p < 0, 0 < q < -\frac{2}{3}p\sqrt{(-\frac{1}{3}p)}$.

CHAPTER 10

EXAMPLES 1

1. $\dfrac{2}{x+1}-\dfrac{2}{x+2}$. **2.** $\dfrac{4}{x-2}-\dfrac{2}{x-1}$.

3. $\dfrac{2}{3(x-1)}+\dfrac{1}{3(x+2)}$. **4.** $\dfrac{1}{x+1}+\dfrac{1}{x+2}$.

5. $\dfrac{3}{x+1}-\dfrac{6}{x+2}+\dfrac{3}{x+3}$. **6.** $\dfrac{3}{x-1}-\dfrac{12}{x-2}+\dfrac{9}{x-3}$.

7. $\dfrac{6}{x-1}-\dfrac{18}{x-2}+\dfrac{12}{x-3}$. **8.** $\dfrac{3}{x+1}-\dfrac{24}{x+2}+\dfrac{27}{x+3}$.

9. $\dfrac{2}{x}-\dfrac{1}{x+2}$. **10.** $\dfrac{2}{x}-\dfrac{1}{x-2}$.

11. $\dfrac{1}{2x}-\dfrac{2}{x-1}+\dfrac{3}{2(x-2)}$. **12.** $\dfrac{5}{6(x+2)}+\dfrac{2}{3(x-1)}-\dfrac{1}{2x}$.

EXAMPLES 2

1. $-\dfrac{1}{(x-1)^2}-\dfrac{1}{x-1}+\dfrac{1}{x-2}$. **2.** $\dfrac{1}{3(x+1)^2}-\dfrac{2}{9(x+1)}+\dfrac{2}{9(x-2)}$.

3. $-\dfrac{1}{x^2}-\dfrac{2}{x}+\dfrac{2}{x-1}$.

4. $-\dfrac{1}{2(x-1)^2}-\dfrac{4}{(x-2)^2}-\dfrac{9}{4(x-1)}+\dfrac{9}{4(x-3)}$.

5. $-\dfrac{18}{(x+2)^3}-\dfrac{15}{(x+2)^2}-\dfrac{14}{x+2}+\dfrac{1}{2(x-1)}+\dfrac{27}{2(x+1)}$.

6. $\dfrac{1}{(x+2)^2}-\dfrac{1}{x+2}+\dfrac{1}{x+1}$. **7.** $-\dfrac{4}{3(x+2)^2}+\dfrac{8}{9(x+2)}+\dfrac{1}{9(x-1)}$.

8. $\dfrac{1}{2(x-1)^3}+\dfrac{5}{4(x-1)^2}+\dfrac{7}{8(x-1)}+\dfrac{1}{8(x+1)}$.

9. $-\dfrac{2}{(x-1)^3}-\dfrac{4}{(x-2)^2}-\dfrac{5}{(x-1)^2}-\dfrac{17}{2(x-1)}+\dfrac{8}{x-2}+\dfrac{1}{2(x-3)}.$

10. $\dfrac{1}{3(x-1)^4}+\dfrac{11}{9(x-1)^3}+\dfrac{43}{27(x-1)^2}+\dfrac{65}{81(x-1)}+\dfrac{16}{81(x+2)}.$

11. $\dfrac{3}{(x-1)^5}+\dfrac{1}{(x-1)^4}.$

12. $1-\dfrac{8}{x+1}+\dfrac{24}{(x+1)^2}-\dfrac{32}{(x+1)^3}+\dfrac{16}{(x+1)^4}.$

EXAMPLES 3

1. $\dfrac{1}{4(x-1)}-\dfrac{x+1}{2(x^2+1)^2}-\dfrac{x+1}{4(x^2+1)}.$

2. $\dfrac{x+1}{(x^2+1)^2}-\dfrac{x+1}{x^2+1}+\dfrac{x+1}{x^2+2}.$

3. $\dfrac{1}{2(x^2-x+1)}+\dfrac{1}{2(x^2+x+1)}.$

4. $\dfrac{1}{2(x^2-x+1)}-\dfrac{1}{2(x^2+x+1)}.$

5. $\dfrac{3}{2(x^2-x+1)}-\dfrac{1}{2(x^2+x+1)}.$

6. $-\dfrac{3}{(x^2+4)^2}-\dfrac{1}{x^2+4}+\dfrac{1}{x^2+1}.$

7. $\dfrac{1}{8(x-1)}-\dfrac{1}{4(x+1)^2}-\dfrac{3}{8(x+1)}+\dfrac{x-1}{4(x^2+1)}.$

8. $\dfrac{1}{32(x-1)}+\dfrac{x+1}{2(x^2+3)^3}-\dfrac{x+1}{8(x^2+3)^2}-\dfrac{x+1}{32(x^2+3)}.$

REVISION EXAMPLES V

1. $a = 1, b = 3, c = \frac{17}{3}, d = -\frac{2}{3}.$

$$x-1+\frac{1}{x+1}-\frac{x-1}{x^2+1}.$$

2. $a = -3, b = -2, c = 9; -\frac{2}{5}.$

3. $\dfrac{1}{x-2}-\dfrac{1}{x-1};\ -\dfrac{1}{(x-1)^2}-\dfrac{3}{x-1}+\dfrac{4}{x-2};\ 1+\dfrac{1}{(x-1)^3}+\dfrac{3}{(x-1)^2}+\dfrac{3}{x-1}.$

4. $\dfrac{1}{3(x-1)}-\dfrac{x+2}{3(x^2+x+1)};\ \dfrac{1}{3(x+1)}-\dfrac{x-2}{3(x^2-x+1)};$

$$\frac{1}{6(x-1)}-\frac{1}{6(x+1)}-\frac{x+2}{6(x^2+x+1)}+\frac{x-2}{6(x^2-x-1)};$$

$$\frac{1}{6(x-1)}+\frac{1}{6(x+1)}-\frac{x+2}{6(x^2+x+1)}-\frac{x-2}{6(x^2-x+1)};$$

$$\frac{1}{6(x-1)}+\frac{1}{6(x+1)}-\frac{x-1}{6(x^2+x+1)}-\frac{x+1}{6(x^2-x+1)};$$

$$\frac{1}{6(x-1)}-\frac{1}{6(x+1)}+\frac{2x+1}{6(x^2+x+1)}-\frac{2x-1}{6(x^2-x+1)}.$$

5. $\dfrac{1}{3(1-x)^2}+\dfrac{1}{3(1-x)}+\dfrac{1+x}{3(1+x+x^2)}.$

6. $\dfrac{\sqrt{2}}{4}\left\{\dfrac{x+\sqrt{2}}{x^2+x\sqrt{2}+1}-\dfrac{x-\sqrt{2}}{x^2-x\sqrt{2}+1}\right\};$

$\dfrac{1}{8(x-1)}-\dfrac{1}{8(x+1)}-\dfrac{1}{4(x^2+1)}-\dfrac{\sqrt{2}}{8}\left\{\dots\right\}.$

7. $\dfrac{1}{4(x-1)^3}+\dfrac{1}{8(x+1)^2}-\dfrac{1}{2(x-1)^2}+\dfrac{11}{16(x-1)}+\dfrac{5}{16(x+1)}-\dfrac{1}{x};$

$\dfrac{1}{x}-\dfrac{x}{(1+x^2)^2}-\dfrac{x}{1+x^2}.$

8. $2x+\dfrac{16}{3(x+2)}+\dfrac{16}{3(x-2)}-\dfrac{1}{3(x+1)}-\dfrac{1}{3(x-1)};$

$-\dfrac{1}{x^3}+\dfrac{1}{x^2}-\dfrac{1}{x}+\dfrac{1}{4(x+1)^2}+\dfrac{9}{8(x+1)}+\dfrac{1}{8(x-1)}-\dfrac{x+1}{4(x^2+1)};$

$\dfrac{2}{3(1-x)}+\dfrac{4+2x}{3(1+x+x^2)}-1;\ \dfrac{1}{2(1-x)^2}+\dfrac{1}{2(1-x)}+\dfrac{x}{2(1+x^2)}.$

9, 10. In each 'solution', the first line is wrong, as the form proposed is not permissible. It should be emphasized that (as the proofs of this chapter indicate) the given expressions must be rational functions, that is, ratios of polynomials.

Chapter 11

EXAMPLES 1

1. (1, 1), (1, 2), ..., (1, 9); (2, 1), (2, 2), (2, 3), (2, 4); (3, 1), (3, 2), (3, 3); (4, 1), (4, 2). For equality, add (1, 10), (2, 5), (5, 1), (5, 2).

2. (i) $-2, -1, 0, 1, 2, 3, 4$; (ii) 3, 4, 5, 6; (iii) 1, 2, 3, 4.

3. (i) 3; (ii) -3; (iii) 5; (iv) 1; (v) 1; (vi) 3;
(vii) 2; (viii) 0; (ix) 16; (x) 5; (xi) 1; (xii) 7.

EXAMPLES 2

1. $x > 2$. **2.** $x > -\frac{3}{2}$. **3.** $x > 0$.

4. All x. **5.** $x > 2, x < 1$. **6.** $x > 3, x < 1$.

7. $x > -2$. **8.** $x > 1$. **9.** $x > 2, x < 0$.

10. $x > -1, -3 < x < -2$. **11.** $|x| > 2$.

12. $x > 2$. **13.** $x > 3, -2 < x < 1$.

14. $x > 2, -3 < x < -1$. **15.** $x > 4, 2 < x < 3, x < 1$.

16. $x > 4, x < 2$.

EXAMPLES 3

1. $k > 1$.
2. $k > 4$.
3. $k > \frac{1}{2}$.
4. $|k| < 1$.
5. $k > 9$.
6. $k > 1$.
7. $|k| < \sqrt{10}$.
8. $k > 16$.
9. $k > 2$ or $k < 1$.
10. All values of k.

REVISION EXAMPLES VI

6. $a > 2\sqrt{6}$.

7. $a > b, a+b > 2c$.

11. $x(x+1)(x+7)$; $x > 0$, $-7 < x < -1$.

12. $\lambda \geqslant 7+4\sqrt{3}$; $\lambda \leqslant 7-4\sqrt{3}$.

21. (i) $x > 2$; (ii) $x > 3$, $1 < x < 2$.

23. $\lambda > 1$.

25. $(ac-b^2)/(ak^2+2bk+c)$.

28. (i) $x > 2$, $x < 1$.
(ii) If $k > 2$, $x > k$, $1 < x < 2$.
If $1 < k < 2$, $x > 2$, $1 < x < k$.
If $k < 1$, $x > 2$, $k < x < 1$.
If $k = 2$, $x > 1$; if $k = 1$, $x > 2$.

29. 4^{-4}.

REVISION EXAMPLES VII

2. If $c > 1$, $y \geqslant 4(c-1)$ or $y \leqslant 0$.
If $c < 1$, $y \leqslant -4(1-c)$ or $y \geqslant 0$.
If $c = 1$, no restriction.

6. Two roots.

8. m positive, 3; m negative, 1.

9. (i) $-15 < a < -3$.
(ii) Let roots of $(1-a)y^2-(9-a)y+9 = 0$ be α, β, real.

Then, if $a > 1$, the function is not between α, β; if $a < 1$, the function is necessarily between α, β.

15. Exclude $1 < y < 2$.

21. Let roots of $ky^2-(ak+c)y+(ac-b^2) = 0$ be α, β, real. Then, if $k > 0$, the function is not between α, β; if $k < 0$, the function is necessarily between α, β.

CHAPTER 13

EXAMPLES 1

1. 120, 120, 280. **2.** 84, 84.

3. 2002, 6, 1876. **4.** $\frac{16}{49}$, $\frac{12}{49}$, $\frac{11}{49}$, $\frac{11}{49}$.

5. $\frac{1}{13}$, $\frac{1}{4}$, $\frac{1}{52}$. **6.** 9!, 2(8!), 7(8!).

7. 4, 10, 56. **8.** $\frac{3}{7}$, $\frac{1}{14}$, $\frac{1}{14}$.

REVISION EXAMPLES VIII

1. 5!; 4!. **2.** 126; 60.

3. 76145. **4.** $\frac{1}{18}$; $\frac{1}{6}$.

5. 59049; 58680. **7.** 9!; 9!/3!.

8. 300. **11.** $\dfrac{13!}{6!\,7!}\times\dfrac{39!}{7!\,32!}$. The second answer is $\dfrac{32!}{6!\,26!}$ (his partner) $\times\dfrac{26!}{13!\,13!}$ (the others) $\div\dfrac{39!}{13!\,13!\,13!}$ (the ways in which his hand remains as given).

17. 14868 (= 6930 + 5775 + 1925 + 231 + 7 for 2, 3, 4, 5, 6 zeros).

19. $\dfrac{(p+1)!}{q!\,(p+1-q)!}$. **21.** $\frac{1}{35}$; $\frac{1}{35}$.

33. $7\times 8!$

35. $\dfrac{(m+n)!}{m!\,n!}$; $\dfrac{(m+n-2)!}{m!\,(n-2)!}$; $\dfrac{n(n-1)}{(m+n)\,(m+n-1)}$.

37. 2624. **39.** $\dfrac{52!}{\{13!\}^4}$.

CHAPTER 14

EXAMPLES 1

1. $x^3+6x^2+11x+6$. **2.** x^3+2x^2-5x-6.

3. $6x^3-11x^2+6x-1$. **4.** $3x^3+4x^2-3x-4$.

5. $x^4+10x^3+35x^2+50x+24$. **6.** $x^4-15x^2+10x+24$.

7. $16x^4-80x^3+140x^2-100x+24$. **8.** $6x^4-5x^3-5x^2+5x-1$.

9. $x^3+(a+2b+3c)\,x^2+(6bc+3ca+2ab)\,x+6abc$.

10. $a^2bx^3+a^3x^2-b^3x-ab^3$.

EXAMPLES 2

1. $x^5-5x^4a+10x^3a^2-10x^2a^3+5xa^4-a^5$.

2. $32x^5+80x^4a+80x^3a^2+40x^2a^3+10xa^4+a^5$.

3. $32x^5+240x^4+720x^3+1080x^2+810x+243$.

4. $32x^5-400x^4+2000x^3-5000x^2+6250x-3125$.

5. $x^3+6x^2a+12xa^2+8a^3$.

6. $81x^4-540x^3a+1350x^2a^2-1500xa^3+625a^4$.

7. $1+6a^2x+15a^4x^2+20a^6x^3+15a^8x^4+6a^{10}x^5+a^{12}x^6$.

8. $a^7x^7-7a^6x^6+21a^5x^5-35a^4x^4+35a^3x^3-21a^2x^2+7ax-1$.

9. $a^3x^3-3a^2bx^2y+3ab^2xy^2-b^3y^3$.

10. $a^5b^5x^5-5a^4b^4x^4y^2+10a^3b^3x^3y^4-10a^2b^2x^2y^6+5abxy^8-y^{10}$.

11. $1-6abx+15a^2b^2x^2-20a^3b^3x^3+15a^4b^4x^4-6a^5b^5x^5+a^6b^6x^6$.

12. $x^{12}-12x^{11}+66x^{10}-220x^9+495x^8-792x^7+924x^6-792x^5+495x^4$
$-220x^3+66x^2-12x+1$.

13. $1024x^{10}+5120x^9+11{,}520x^8+15{,}360x^7+13{,}440x^6+8064x^5$
$+3360x^4+960x^3+180x^2+20x+1$.

14. $x^9-18x^8y+144x^7y^2-672x^6y^3+2016x^5y^4-4032x^4y^5$
$+5376x^3y^6-4608x^2y^7+2304xy^8-512y^9$.

15. $16x^4-96ax^3y+216a^2x^2y^2-216a^3xy^3+81a^4y^4$.

16. $x^{14}+7ax^{12}y+21a^2x^{10}y^2+35a^3x^8y^3+35a^4x^6y^4+21a^5x^4y^5+7a^6x^2y^6+a^7y^7$.

REVISION EXAMPLES IX

1. $\pm\sqrt{(14/3)}$.
2. -74.
3. 0·00054.
4. 1·0829.
5. (2, 1), (−2, −1).
7. $1+8x+36x^2+104x^3$.
8. −161.
9. 0·86813.
10. 3, 2, 6.
11. $y^7-7y^5+14y^3-7y$.
12. $y^6-5y^4+6y^2-1$.
13. 5, −4.
14. 0·9231.
15. $\dfrac{14!\,2^7}{(7!)^2}$; $\dfrac{14!\,3^{12}}{12!\,2!}\left(\dfrac{2}{3}\right)^2$.
16. $\dfrac{(-)^r\,n!}{r!\,(n-r)!}$; 35.

17. $\dfrac{13!}{5!\,8!}\left(\dfrac{1}{3}\right)^5\left(\dfrac{1}{2}\right)^8.$ **19.** 1 051.

20. $\frac{1}{15}$. **26.** $m = 3,\ n = 2.$

28. Second last. **31.** 11. **32.** 9, 10, 11, 12, 13, 14.

CHAPTER 15

Warning example. Anything. For example, if the pth term were

$$(p-1)(p-2)(p-3)(p-4)+1,$$

the next term would be 25. It is essential not to be misled by pattern, though that can, indeed, form a valuable intuitive guide.

REVISION EXAMPLES X

1. 25. **2.** 426. **3.** $\frac{9}{4}$, 3, $\frac{15}{4}$.

4. 26th; $-8\frac{1}{3}$. **5.** 11,571. **7.** $\frac{1}{2}n(3n+1)$; 7.

8. -1, 3, 7, 11. **9.** 30th; $\frac{1}{3}n(58-n)$; 59.

10. 47,500; 16,167. **11.** $\frac{1}{2}(c-a)$; $9(2c-a)$. **12.** 40,400.

13. 2675; 27. **14.** 1732. **15.** $-\frac{3}{2}$; -65.

17. 20. **19.** 35. **20.** 21.

22. $6\frac{1}{2}$. **24.** 6417. **25.** 5, 3; 36.

26. $-n$, $n+1$. **27.** $n(3n-4)$. **28.** 67.

29. $-\frac{1}{2}$, $\frac{1}{2}$, $\frac{3}{2}$, $\frac{5}{2}$. **31.** 1,989,924; 371. **32.** 2, 6, 18.

33. 4, -12; $15\frac{7}{8}$, $57\frac{7}{8}$. **34.** 20. **36.** $\frac{3}{2}$, $\frac{729}{8}$.

37. $3^{11}-1$. **38.** $\dfrac{a(1-k^{13})}{k^6(1-k)}$. **40.** 14.

41. (i) $2p^3-3pq+r=0$; (ii) $p^3r=q^3$; (iii) $2q^3-3pqr+r^2=0$.

43. $-\frac{5}{3}$, $-\frac{2}{3}$, $\frac{1}{3}$, $\frac{4}{3}$. **45.** $\dfrac{x(1-x^n)}{(1-x)^2}-\dfrac{nx^{n+1}}{1-x}.$

46. $\frac{1}{8}(2n-1)(2n+1)(2n+3)(2n+5)+\frac{15}{8}$.

47. 13. **48.** $\frac{1}{2}n^2(n+1)$.

49. (i) $\dfrac{3}{4}-\dfrac{(2n+3)}{2(n+1)(n+2)}$; (ii) $\dfrac{n}{3n+1}$.

50. $\dfrac{1-(2n-1)x^n}{1-x}+\dfrac{2x(1-x^{n-1})}{(1-x)^2}.$ **51.** $\frac{1}{6}n(n+1)(2n+1)$; 6.

56. $\frac{1}{12}n(n+1)(n+2)(3n+17)$.

58. $\dfrac{4}{x}-\dfrac{7}{x+3}+\dfrac{3}{x+5}$;

$\dfrac{359}{360}-\dfrac{2}{3}\left(\dfrac{1}{n+1}+\dfrac{1}{n+2}+\dfrac{1}{n+3}\right)+\dfrac{1}{2}\left(\dfrac{1}{n+4}+\dfrac{1}{n+5}\right)$.

59. $\dfrac{1}{60}-\dfrac{1}{6(3n+2)(3n+5)}$.

60. (i) $\frac{1}{6}n(n+1)(2n+7)$, (ii) $\dfrac{x(1-x^n)}{(1-x)^2}+\dfrac{2x-(n+2)x^{n+1}}{1-x}$.

61. $\dfrac{2n}{n+1}$. **62.** $\dfrac{5}{4}-\dfrac{2n+5}{2(n+1)(n+2)}$. **63.** $\dfrac{n}{5n+1}$.

64. $n-6+\dfrac{12(3n+10)}{(n+4)(n+5)}$. **65.** $-31{,}940$.

66. (i) $\dfrac{n(n+1)}{2(2n+1)(2n+3)}$; (ii) $\dfrac{n(8n-1)}{21(4n+3)(4n+7)}$.

CHAPTER 16

EXAMPLES 1

1. 6. **2.** 12. **3.** 7.

4. (i) 3, (ii) 4, (iii) 5. **5.** (i) 10, (ii) 11, (iii) 14.

EXAMPLES 2

5. $\dfrac{3}{4}-\dfrac{4n+3}{2(n+1)(n+2)}\to\dfrac{3}{4}$.

6. $(S-S_n)/S < 10^{-6}$ when $n > 1414$. **7.** 1.

REVISION EXAMPLES XI

2. $-\frac{7235}{1296}$. **3.** $1+x^4+\frac{5}{8}x^8+\frac{5}{16}x^{12}$.

4. $1-x-\frac{1}{2}x^2-\frac{1}{2}x^3-\frac{5}{8}x^4$, $1-x-\frac{1}{2}x^2-\frac{1}{4}x^3-\frac{1}{8}x^4$;
$-\frac{1}{2} < x < \frac{1}{2}$; $-2 < x < 2$.

5. 0·332230. **6.** $1+\frac{2}{5}x+\frac{12}{25}x^2+\frac{88}{125}x^3$.

7. 2·009926. **8.** $(r+1)(2r+1)$. **9.** 0·751.

10. $\frac{1}{2}$; $\frac{1}{8}x^4+\frac{1}{8}x^6+\frac{15}{128}x^8$. **11.** $1-\frac{1}{2}x+\frac{3}{8}x^2-\frac{5}{16}x^3+\frac{35}{128}x^4$.

12. 1, 4, 1. **13.** 1·070344. **14.** 9·949874.

17. $n>\dfrac{(k-1)x}{1-x}$; 20th and 21st; 1st. **19.** 0, 0, 0, 0, $-\frac{3}{2}$.

REVISION EXAMPLES XII

1. $\dfrac{3}{x+1}-\dfrac{12}{(x+1)^2}+\dfrac{18}{(x+1)^3}-\dfrac{1}{(x+1)^5}$.

2. $\dfrac{1}{2(x-1)}+\dfrac{1}{162(x+1)}+\dfrac{2}{3(x-2)^4}-\dfrac{5}{9(x-2)^3}+\dfrac{14}{27(x-2)^2}-\dfrac{41}{81(x-2)}$.

4. $\dfrac{5}{9(x-2)}+\dfrac{1}{3(x+1)^2}-\dfrac{5}{9(x+1)}$.

5. $a=\frac{3}{2}$, $b=\frac{1}{2}$, $c=-\frac{1}{2}$.

6. $\dfrac{2}{3(x-1)^2}+\dfrac{4}{9(x-1)}+\dfrac{5}{9(x+2)}$; $\frac{1}{2}+\frac{3}{4}x+\frac{13}{8}x^2+\frac{35}{16}x^3$; $|x|<1$.

7. $\dfrac{1}{x-1}+\dfrac{3x+2}{x^2+2}$; $\frac{1}{2}x-\frac{3}{2}x^2-\frac{7}{4}x^3-\frac{3}{4}x^4$; $|x|<1$.

8. $\dfrac{4}{5(1+2x)}+\dfrac{(1-2x)}{5(1+x^2)}$; (i) 51, (ii) -410.

9. $\dfrac{1}{2(x-1)}-\dfrac{2}{x-2}+\dfrac{3}{2(x-3)}$; $\dfrac{2}{(x-2)^3}+\dfrac{1}{(x-2)^2}$;

$-\dfrac{1}{2}+\dfrac{1}{2^{n+1}}-\dfrac{1}{2.3^{n+1}}$; $|x|<3$.

10. $\dfrac{1}{x-1}-\dfrac{3}{(x+2)^2}-\dfrac{1}{x+2}$; $-1-(-\frac{1}{2})^{n+2}(3n+5)$.

11. $\dfrac{2}{(x-2)^2}-\dfrac{1}{x-2}+\dfrac{x+1}{x^2+x+4}$; $\frac{5}{4}+\frac{15}{16}x+\frac{25}{64}x^2+\frac{75}{256}x^3+\frac{225}{1024}x^4$.

14. $a=0$; $\dfrac{19}{18}+\dfrac{1}{3(N+1)}-\dfrac{5}{3(N+2)}-\dfrac{5}{3(N+3)}$.

15. $-\dfrac{7}{6(x-1)^2}+\dfrac{65}{36(x-1)}+\dfrac{5}{4(x-3)}-\dfrac{(5x-7)}{9(x^2+2)}$;

$-\dfrac{84n-23}{36}-\dfrac{5}{12.3^{2n}}+\dfrac{7}{18.(-2)^n}$; $-\dfrac{8n+19}{36}-\dfrac{5}{12.3^{2n+1}}-\dfrac{5}{18.(-2)^n}$.

16. $\dfrac{2}{(x+1)^4}-\dfrac{2}{(x+1)^2}+\dfrac{1}{x+1}$.

17. $\dfrac{1}{(x-1)^3}+\dfrac{1}{(x-1)^2}+\dfrac{5}{(x-2)^3}-\dfrac{1}{(x-2)^2}$; $|x|<1$.

18. (i) $\frac{(-)^{n+1}2}{3^n(x+2)}+\frac{1}{3(x-1)^n}+\frac{2}{3^2(x-1)^{n-1}}-\frac{2}{3^3(x-1)^{n-2}}+\ldots+\frac{(-)^n\,2}{3^n(x-1)}$;

(ii) $\frac{(-1)^n}{x-1}-\frac{(-2)^n}{3^n(2x-1)}+\frac{2}{3(x-2)^n}-\frac{1-\frac{2}{3^2}}{(x-2)^{n-1}}+\frac{1-\frac{2^2}{3^3}}{(x-2)^{n-2}}-\ldots$

$$+(-)^{n-1}\frac{1-\frac{2^{n-1}}{3^n}}{x-2}.$$

19. $\frac{(-1)^{n+1}}{2x}+\frac{1}{2(x-2)}-\frac{1}{(x-1)^n}-\frac{1}{(x-1)^{n-2}}-\frac{1}{(x-1)^{n-4}}-\ldots$, the last term being $\frac{-1}{(x-1)^2}$ if n is even and $\frac{-1}{x-1}$ if n is odd.

20. $\frac{2}{(x-1)^3}+\frac{2}{(x-1)^2}-\frac{1}{x-1}+\frac{x-1}{x^2+1}$.

21. $\frac{1}{(x-a)^n}-\frac{(2a)^{-1}n}{(x-a)^{n-1}}+\frac{(2a)^{-2}\,n(n+1)}{2!\,(x-a)^{n-2}}-\ldots$

$$+(-)^{n-1}\frac{(2a)^{-(n-1)}\,n(n+1)\ldots(2n-2)}{(n-1)!\,(x-a)}$$

$$+(-)^n\left\{\frac{1}{(x+a)^n}+\frac{(2a)^{-1}n}{(x+a)^{n-1}}+\frac{(2a)^{-2}\,n(n+1)}{2!\,(x+a)^{n-2}}+\ldots\right.$$

$$\left.+\frac{(2a)^{-(n-1)}\,n(n+1)\ldots(2n-2)}{(n-1)!\,(x+a)}\right\}.$$

CHAPTER 18

EXAMPLES 1

1. $1+2x+2x^2+\frac{4}{3}x^3+\ldots+\frac{2^{k-1}}{(k-1)!}x^{k-1}+\ldots$.

2. $1-2x+2x^2-\frac{4}{3}x^3+\ldots+(-)^{k-1}\frac{2^{k-1}}{(k-1)!}x^{k-1}+\ldots$.

3. $1+5x+\frac{25}{2}x^2+\frac{125}{6}x^3+\ldots+\frac{5^{k-1}}{(k-1)!}x^{k-1}+\ldots$.

4. $1-x^2+\frac{1}{2}x^4-\frac{1}{6}x^6+\ldots+(-)^{k-1}\frac{1}{(k-1)!}x^{2k-2}+\ldots$.

5. $1+3y+\frac{9}{2}y^2+\frac{9}{2}y^3+\ldots+\frac{3^{k-1}}{(k-1)!}y^{k-1}+\ldots$.

6. $1-4a+8a^2-\frac{32}{3}a^4+\ldots+(-)^{k-1}\dfrac{4^{k-1}}{(k-1)!}a^{k-1}+\ldots.$

7. $1+2x+\frac{3}{2}x^2+\frac{2}{3}x^3+\ldots+\dfrac{k}{(k-1)!}x^{k-1}+\ldots.$

8. $1-2x+\frac{3}{2}x^2-\frac{2}{3}x^3+\ldots+(-)^{k-1}\dfrac{k}{(k-1)!}x^{k-1}+\ldots.$

9. $2+3x+\frac{5}{2}x^2+\frac{3}{2}x^3+\ldots+\dfrac{1+2^{k-1}}{(k-1)!}x^{k-1}+\ldots.$

11. (i) e; (ii) $2e-1$; (iii) $e+1$.

12. (i) $1+x+\frac{3}{2}x^2+\frac{7}{6}x^3$.

(ii) $1-x+\frac{3}{2}x^2-\frac{7}{6}x^3$.

(iii) $1+x+\frac{3}{2}x^3+\frac{13}{6}x^3$.

REVISION EXAMPLES XIII

1. $\dfrac{-(n-1)(n-2)(n-3)}{3!\,n!}$.

2. $a=\frac{1}{2}$, $b=\frac{1}{10}$, $c=\frac{1}{120}$.

3. $468{\cdot}3$ $(=(8{\cdot}1)^8\div(8!))$.

4. $a=4$, $b=-5$, $c=1$.

5. $n<x<\sqrt{\{n(n+1)\}}$.

6. $\dfrac{(-2)^{n-2}}{n!}\{4-2np+n(n-1)q\}$.

7. $1+x-\frac{1}{5}x^5$.

8. $\dfrac{(-)^{r-1}(r-1)\,3^r}{r!}$.

9. $15e$.

10. $\left(1-\dfrac{3}{x}+\dfrac{3}{x^2}\right)e^x+\frac{1}{2}-\dfrac{3}{x^2}$.

CHAPTER 19

EXAMPLES 1

1. (i) $2x-\frac{4}{2}x^2+\frac{8}{3}x^3-\ldots+(-)^{r-1}\dfrac{2^r}{r}x^r+\ldots;\ -\frac{1}{2}<x\leqslant\frac{1}{2}$.

(ii) $-3x-\frac{9}{2}x^2-\frac{27}{3}x^3-\ldots-\dfrac{3^r}{r}x^r-\ldots;\ -\frac{1}{3}\leqslant x<\frac{1}{3}$.

(iii) $x^2-\frac{1}{2}x^4+\frac{1}{3}x^6-\ldots+(-)^{r-1}\dfrac{1}{r}x^{2r}+\ldots;\ -1\leqslant x\leqslant 1$.

(iv) $3x-\frac{5}{2}x^2+3x^3-\frac{17}{4}x^4+\ldots+(-)^{r-1}\dfrac{1+2^r}{r}x^r+\ldots;\ -\frac{1}{2}<x\leqslant\frac{1}{2}$.

(v) $2x+\frac{2}{3}x^3+\frac{2}{5}x^5+\ldots+\dfrac{2}{2r+1}\,x^{2r+1}+\ldots;\ -1 < x < 1.$

(vi) $x-\frac{5}{2}x^2+\frac{19}{3}x^3-\frac{65}{4}x^4+\ldots+(-)^{r-1}\,\dfrac{3^r-2^r}{r}\,x^r+\ldots;\ -\frac{1}{3} < x \leqslant \frac{1}{3}.$

2. (i) $\log_e 2+\frac{3}{2}x-\frac{1}{2}(\frac{3}{2}x)^2+\frac{1}{3}(\frac{3}{2}x)^3-\frac{1}{4}(\frac{3}{2}x)^4+\ldots.$

(ii) $\log_e 2-\frac{3}{2}x-\frac{1}{2}(\frac{3}{2}x)^2-\frac{1}{3}(\frac{3}{2}x)^3-\frac{1}{4}(\frac{3}{2}x)^4-\ldots.$

(iii) $\log_e 9-\frac{4}{9}x^2-\frac{1}{2}(\frac{4}{9}x^2)^2-\frac{1}{3}(\frac{4}{9}x^2)^3-\frac{1}{4}(\frac{4}{9}x^2)^4-\ldots.$

(iv) $2\{\frac{1}{2}x+\frac{1}{3}(\frac{1}{2}x)^3+\frac{1}{5}(\frac{1}{2}x)^5+\frac{1}{7}(\frac{1}{2}x)^7+\ldots\}.$

(v) $-5x-\frac{17}{2}x^2-\frac{65}{3}x^3-\ldots-\dfrac{1+4^r}{r}\,x^r-\ldots.$

(vi) $\log_e 6+(\frac{2}{3}-\frac{3}{2})\,x-\frac{1}{2}(\frac{4}{9}+\frac{9}{4})\,x^2+\frac{1}{3}(\frac{8}{27}-\frac{27}{8})\,x^3-\frac{1}{4}(\frac{16}{81}+\frac{81}{16})\,x^4+\ldots.$

REVISION EXAMPLES XIV

1. $\log_e 2+\frac{3}{2}x+\frac{3}{8}x^2+\frac{3}{8}x^3+\ldots+\dfrac{1}{r}\{1-(-\frac{1}{2})^r\}x^r+\ldots.$

2. $-\frac{5}{6}x-\frac{1}{2}(\frac{5}{6})^2x^2-\frac{1}{3}(\frac{5}{6})^3x^3-\ldots-\dfrac{1}{r}(\frac{5}{6})^r x^r-\ldots,$ if $-\frac{6}{5} \leqslant x < \frac{6}{5}$;

$-\log 6+(\frac{1}{2}+\frac{1}{3})\,x+\dfrac{1}{2}\left(\dfrac{1}{2^2}+\dfrac{1}{3^2}\right)x^2+\ldots+\dfrac{1}{r}\left(\dfrac{1}{2^r}+\dfrac{1}{3^r}\right)x^r+\ldots,$ if $-2 \leqslant x < 2.$

4. $-x-\frac{5}{2}x^2-\frac{7}{3}x^3;\ \dfrac{1}{r}\{(-1)^{r-1}-2^r\};\ -\frac{1}{2} \leqslant x < \frac{1}{2}.$

5. $-3x^2-10x^3-\frac{57}{2}x^4;\ \dfrac{1}{r}\{3.2^r-2.3^r\};\ -\frac{1}{3} \leqslant x < \frac{1}{3}.$

6. $\dfrac{(-)^r\,9.6^{r-2}(r-1)(r-4)}{r!}.$

7. 2·1972.

8. $2\left\{\left(\dfrac{1}{2n+1}\right)+\dfrac{1}{3}\left(\dfrac{1}{2n+1}\right)^3+\dfrac{1}{5}\left(\dfrac{1}{2n+1}\right)^5+\ldots\right\}.$

9. 0·223143.

10. 0·4771.

12. $y+\frac{1}{2}y^2+\frac{1}{3}y^3+\ldots.$

13. 0·6931.

14. $\log_e 2+\frac{5}{2}x+\frac{7}{8}x^2+\frac{17}{24}x^3+\ldots;\ -1 \leqslant x < 1.$

15. $b = 0{\cdot}04879$, $c = -0{\cdot}02020$; 0·0690, 0·1178.

16. 2·3026.

18. $\dfrac{1}{4(1-x)}+\dfrac{3}{(2+x)^2}.$

21. 3, 4.

22. $\frac{68}{3}x^3.$

23. $-\frac{19}{216}x^4$.

26. $\frac{5}{2}$; $|x| < \frac{1}{2}$

27. $a = b = \frac{1}{2}$.

28. $a = \frac{2}{3}$; $b = \frac{1}{6}$.

31. $2x + \dfrac{2x^3}{3!} + \dfrac{2x^5}{5!} + \ldots$.

CHAPTER 20

EXAMPLES 1

1. 1. **2.** 44. **3.** 1. **4.** $a-b$.

5. a^2-b^2. **6.** 0. **7.** 9. **8.** 63.

9. -409. **10.** -144. **11.** 0. **12.** 0.

13. $a^3+b^3+c^3-3abc$.

14. $abc+2fgh-af^2-bg^2-ch^2$.

15. 39. **16.** 57. **17.** 0.

18. $a_1b_2c_3-a_1b_3c_2+a_2b_3c_1-a_2b_1c_3+a_3b_1c_2-a_3b_2c_1$.

EXAMPLES 2

1. $\lambda\mu\nu = -1$.

2. $A = 7$, $B = 0$.

EXAMPLES 3

2. 0. **3.** 0. **4.** -384. **5.** 362.

6. 806. **7.** -4516. **8.** 2. **9.** 127.

EXAMPLES 4

2. $k = 2$.

3. $k = 0$.

6. (i) $a \neq 1$, unique; $a = 1$, $b \neq 3$, no solution; $a = 1$, $b = 3$, solution $(\lambda, 3-\lambda)$.

(ii) $a \neq \frac{1}{2}$, unique; $a = \frac{1}{2}$, $b \neq \frac{5}{2}$, no solution; $a = \frac{1}{2}$, $b = \frac{5}{2}$, solution $(\lambda, 5-2\lambda)$.

(iii) $a \neq -\frac{2}{3}$, unique; $a = -\frac{2}{3}$, $b \neq -\frac{8}{3}$, no solution; $a = -\frac{2}{3}$, $b = -\frac{8}{3}$, solution $\left(\lambda, \dfrac{\lambda-4}{3}\right)$.

7. $k = -6$, with solution $(-3, -1)$; $k = 3$, with solution $(3-3\lambda, \lambda)$.

8. $k = -2$, with solution $(-\frac{1}{2}, 1)$; $k = 1$, with solution $(\lambda, 2\lambda-1)$.

EXAMPLES 5

1. $(\frac{1}{2}b, -b, \frac{1}{2}b)$; when $b = 0$, $(\lambda, -2\lambda, \lambda)$.

2. $\left(\frac{b}{a}, -\frac{2b}{a}, \frac{b}{a}\right)$; $a = 0$, $b \neq 0$, no solution; $a = 0$, $b = 0$, $(\lambda, -2\lambda, \lambda)$.

3. $x = \frac{b-1}{3(a+17)}$, $y = -\frac{5a+b+84}{12(a+17)}$, $z = -\frac{a-11b+28}{12(a+17)}$; $a = -17$, $b \neq 1$, no solution; $a = -17$, $b = 1$, $\left(-\frac{12\lambda+5}{3}, \lambda, -\frac{33\lambda+14}{3}\right)$.

4. $a \neq 2$, solution $(0, \frac{2}{3}, \frac{7}{3})$; $a = 2$, solution $(2+3\lambda, -\lambda, 1-2\lambda)$.

5. $x = \frac{b-5}{a-3}$, $y = \frac{a+b-8}{2(a-3)}$, $z = \frac{3a-5b+16}{2(a-3)}$; $a = 3$, $b \neq 5$, no solution; $a = 3$, $b = 5$, $(2\lambda-1, \lambda, 4-5\lambda)$.

6. $x = \frac{2a-5b-12}{4(a-1)}$, $y = \frac{b+2a}{4(a-1)}$, $z = \frac{b+2}{a-1}$; $a = 1$, $b \neq -2$, no solution; $a = 1$, $b = -2$, $(3-5\lambda, \lambda, -2+4\lambda)$.

7. $a = -1$, no solution; $a = 1$, $(1, \frac{1}{2}+\lambda, \frac{1}{2}-\lambda)$; otherwise, $\left(1, \frac{1}{1+a}, \frac{1}{1+a}\right)$.

8. $a = -2$, need $p+q+r = 0$, solution $\left(\lambda, \lambda-\frac{2q+r}{3}, \lambda-\frac{q+2r}{3}\right)$; $a = 1$, need $p = q = r$; solution $(\lambda, \mu, p-\lambda-\mu)$; otherwise $x = \frac{(a+1)p-q-r}{(a-1)(a+2)}$, etc.

9. $x = -\frac{(b-1)^2}{2(a-3)}$, $y = -\frac{(b-1)^2}{2(a-3)}-\frac{1}{2}(1-3b^2)$, $z = 1-2b^2+\frac{3(b-1)^2}{2(a-3)}$; if $a = 3$, $b \neq 1$, no solution; if $a = 3$, $b = 1$, $(\lambda, \lambda+1, -3\lambda-1)$.

10. $x = -3$, $y = \frac{3}{1-k}$, $z = \frac{1-4k}{1-k}$; if $k = 0$, $(\lambda, -\lambda, -2-\lambda)$; if $k = 1$, no solution.

11. $x = \frac{a^2-(b-c)^2}{4bc}$, etc.; if $a = 0$, $b \neq c$, no solution, but if $a = 0$, $b = c$, solution $(0, \lambda, 1-\lambda)$, with cyclic interchange on roles of a, b, c; if $b=c=0$, no solution.

12. $x = y = z = (a+b+c)^{-1}$; if $a+b+c = 0$, no solution; if $a = b = c$, solution $x = \lambda$, $y = \mu$, $z = \frac{1}{a}-\lambda-\mu$ unless $a = 0$, when no solution.

13. $x = \frac{1+b+c}{1+a+b+c}$, $y = z = \frac{-a}{1+a+b+c}$, so that, if $a+b+c = 0$, $x = 1-a$, $y = -a$, $z = -a$; if $1+a+b+c = 0$, no solution unless $a = 0$ also, in which case, $(1+\lambda, \lambda, \lambda)$.

14. $x = \dfrac{(c-2)(2-b)(2+b+c)}{(c-a)(a-b)(a+b+c)}$, etc.; if $a+b+c = 0$, no solution in general; if $a+b+c = 0$, $a = 2$, solution $x = 1+(b-c)\lambda$, $y = (c-2)\lambda$, $z = -(b-2)\lambda$, with cyclic interchange.

15. $\left(\dfrac{1-b}{a(1-ab)}, \dfrac{b(1-a^2)}{1-ab}, \dfrac{a^2-a}{1-ab}\right)$; if $a = 0$, $b \neq 1$, no solution; if $a = 0$, $b = 1$, $(\lambda, 1, 0)$; if $ab = 1$, $b \neq 1$, no solution; if $a = b = 1$, $(\lambda, \mu, 1-\lambda-\mu)$.

16. $a = 1$, $(1+\lambda, 1-\lambda, 1)$; $a = 2$, $(\frac{5}{4}, \frac{1}{2}, \frac{3}{4})$.

17. $\lambda = 1$, $(1, 1, -2)$; $\lambda = -1$, $(1, -1, 0)$; $\lambda = 4$, $(1, 1, 1)$.

18. $\lambda = 2$, $(1, -1, 0)$; $\lambda = 6$, $(1, 1, -1)$; $\lambda = -6$, $(1, 1, 2)$.

20. $x = (lp-mc+nb)/(l^2+m^2+n^2)$, etc.

EXAMPLES 6

1. 5. **2.** -3. **3.** $\frac{5}{2}$.

4. $-1, 2$. **5.** $a, 1$. **6.** a, b.

7. $1, 2, -3$. **8.** $a, 2, -(a+2)$. **9.** $a, b, -(a+b)$.

10. $2, 3, -\frac{6}{5}$. **11.** $a, b, -\dfrac{ab}{a+b}$.

EXAMPLES 7

1. $(b-c)(c-a)(a-b)$. **2.** $(b-c)(c-a)(a-b)$.

3. $abc(b-c)(c-a)(a-b)$. **4.** $-(b-c)(c-a)(a-b)(a+b+c)$.

EXAMPLES 8

4. $\begin{vmatrix} b & c & a \\ a & b & c \\ 0 & 0 & 0 \end{vmatrix}$. Note the consequence that $\Delta = 0$.

5. $$\begin{vmatrix} 1 & 1 & 1 \\ \alpha^4 & \beta^4 & \gamma^4 \\ \alpha^6 & \beta^6 & \gamma^6 \end{vmatrix} \times \begin{vmatrix} 1 & \alpha^2 & \alpha^3 \\ 1 & \beta^2 & \beta^3 \\ 1 & \gamma^2 & \gamma^3 \end{vmatrix}.$$

7. $$\begin{vmatrix} y & x & 1 \\ x & 1 & y \\ 1 & y & x \end{vmatrix}.$$

REVISION EXAMPLES XV

1. (i) 0, (ii) 0, (iii) 0, (iv) -2512.

2. (i) $-44{,}388$. (ii) 1936. (iii) 960. (iv) 5220. (v) -896.

11. $a-b,\ a-b,\ a+2b$. **13.** $-\cos\theta$.

16. 2, 3, -5. **17.** 3, 5, $-\frac{3}{2}$. **19.** 0, a, b, $-(a+b)$.

21. (i) $2(b-c)(c-a)(a-b)(a+b+c)$;
(ii) $-(b-c)(c-a)(a-b)(a+b+c)^2$.

25. $x^3+x(a^2+b^2+c^2)$. **26.** 2, 3 or 3, 2.

27. 1, a, $-\dfrac{a}{1+a}$. **28.** 5, 5, -1.

31. (i) $-(y-z)(z-x)(x-y)(x+y+z)^2$.
(ii) $(y-z)(z-x)(x-y)(x+y+z)(x^3+y^3+z^3)$.

32. $-(a+b+c)(-a+b+c)(a-b+c)(a+b-c)$.

33. 2, 3, -5, 0. **34.** $b, c, d, -(b+c+d)$.

35. 0, 0, 0, $a+b+c+d$.

MISCELLANEOUS EXAMPLES

3. $a=-c,\ b=-c$; or $a=-c,\ b=c$; or $a=c,\ b=-c$.

5. $1\pm\sqrt{2},\ 1\pm\sqrt{3}$. **6.** 1, 5, $2\pm a$. **7.** $k=-3$.

8. End points 0, $4(\beta-\alpha)$. **10.** $1-1/(n+1)^2$.

12. In the general case, $x=\dfrac{bc}{abc+bc+ca+ab}$, etc.

13. $\frac{1}{2}Q^2,\ -\frac{5}{6}QR$.

14. $1+\dfrac{1-x}{x}\log(1-x),\ \dfrac{3}{4}-\dfrac{1}{2x}-\dfrac{(1-x)^2}{2x^2}\log(1-x)$;

$\dfrac{1}{4}-\dfrac{1}{2(N+1)}+\dfrac{1}{2(N+2)}$.

15. 2, $\frac{1}{2}$, 4, $\frac{1}{4}$, $\pm i$. **17.** $a=\pm\frac{1}{3}\sqrt{3},\ b=\mp\frac{1}{9}\sqrt{3}$.

19. $(-1)^{r-1}$. **20.** $-2x^{-1}\log(1-x)$.

21. $2p^3-9pq+27r=0$. **26.** $27p^3$.

27. $-(1-x)(x-y)(1-xy)$.

28. $a+b+c,\ a+\omega b+\omega^2c,\ a+\omega^2b+\omega c$.

30. $2p^3r$. **32.** $0, \dfrac{\pi}{10}, \dfrac{3\pi}{10}, \dfrac{\pi}{8}, \dfrac{3\pi}{8}, \dfrac{\pi}{6}, \dfrac{\pi}{2}$.

34. $(x+a+b)(x+a-b)\{(x-a)^2+b^2\}$

35. $(af-be+cd)^2$. **36.** 0, 0, 1, 1, 1, 1.